Food For Change

Anthropology, Culture and Society

Series Editors:
Professor Vered Amit, Concordia University
and
Professor Christina Garsten, Stockholm University

Recent titles:

Food For Change

The Politics and Values
of Social Movements

Jeff Pratt and Pete Luetchford

with Myriem Naji and Sara Avanzino

PlutoPress
www.plutobooks.com

First published 2014 by Pluto Press
345 Archway Road, London N6 5AA

www.plutobooks.com

Distributed in the United States of America exclusively by
Palgrave Macmillan, a division of St. Martin's Press LLC,
175 Fifth Avenue, New York, NY 10010

British Library Cataloguing in Publication Data
A catalogue record for this book is available from the British Library

ISBN 978 0 7453 3449 3 Hardback
ISBN 978 0 7453 3448 6 Paperback
ISBN 978 1 7837 1000 3 PDF eBook
ISBN 978 1 7837 1005 8 Kindle eBook
ISBN 978 1 7837 1004 1 EPUB eBook

Library of Congress Cataloging in Publication Data applied for

10 9 8 7 6 5 4 3 2 1

Typeset from disk by Stanford DTP Services, Northampton, England
Simultaneously printed digitally by CPI Antony Rowe, Chippenham, UK and
Edwards Bros in the United States of America

Contents

Series Preface

Anthropology is a discipline based upon in-depth ethnographic works that deal with wider theoretical issues in the context of particular, local conditions – to paraphrase an important volume from the series: *large issues* explored in *small places*. This series has a particular mission: to publish work that moves away from an old-style descriptive ethnography that is strongly area-studies oriented, and offer genuine theoretical arguments that are of interest to a much wider readership, but which are nevertheless located and grounded in solid ethnographic research. If anthropology is to argue itself a place in the contemporary intellectual world, then it must surely be through such research.

We start from the question: 'What can this ethnographic material tell us about the bigger theoretical issues that concern the social sciences?' rather than 'What can these theoretical ideas tell us about the ethnographic context?' Put this way round, such work becomes *about* large issues, *set in* a (relatively) small place, rather than detailed description of a small place for its own sake. As Clifford Geertz once said, 'Anthropologists don't study villages; they study *in* villages.'

By place, we mean not only geographical locale, but also other types of 'place' – within political, economic, religious or other social systems. We therefore publish work based on ethnography within political and religious movements, occupational or class groups, among youth, development agencies, and nationalist movements; but also work that is more thematically based – on kinship, landscape, the state, violence, corruption, the self. The series publishes four kinds of volume: ethnographic monographs; comparative texts; edited collections; and shorter, polemical essays.

We publish work from all traditions of anthropology, and all parts of the world, which combines theoretical debate with empirical evidence to demonstrate anthropology's unique position in contemporary scholarship and the contemporary world.

Professor Vered Amit
Professor Christina Garsten

Preface

This study of alternative food politics and values, framed as a response to mainstream provision, is the outcome of long conversations and close collaborations. The approach reflects the authors' common interests in economic and political anthropology, but every effort has been made to render the material and the arguments accessible to anyone concerned at the direction taken by industrialised food, and the challenge of building viable alternatives.

The authors met in 1996 in the Department of Anthropology at the University of Sussex, where they delivered courses together in political anthropology. Jeff Pratt joined the department in 1976 after doctoral and postdoctoral research in a number of locations in Tuscany. Over the years he has continued to conduct fieldwork in Italy, and has published widely on rural transformations and political movements in Europe. More recently he has turned to the study of food, its values and politics. Pete Luetchford joined Jeff Pratt at Sussex when he returned to the UK after several years in Spain, to study for his PhD. His doctoral work, later published as *Fair Trade and a Global Commodity: Coffee in Costa Rica* (Luetchford 2008a), combined his interests in the problems and possibilities of building alternative economies with a concern with ethical ideas informing exchange.

These strands of rural transformations, political movements, and alternative economic practices and values slowly coalesced as the authors engaged in discussions on recent scholarship in economic anthropology. Particularly significant was their close reading of David Graeber's innovative work *Towards an Anthropological Theory of Value* (Graeber 2001), and its arguments about the importance of actions in constituting value. From there, talk turned to the relevance for contemporary anthropology of older writings by Karl Marx, Marcel Mauss and Karl Polanyi, and to specific issues such as the just price, political economy, ethical consumption, alienation and authenticity. Out of this work other collaborations grew. A conference was organised with our colleague Geert de Neve on the theme of fair trade, ethical consumption and corporate social responsibility, published as *Hidden*

Hands in the Market (de Neve, Luetchford, Pratt & Wood 2008). Later, Pete Luetchford worked with James Carrier to produce the volume *Ethical Consumption* (Carrier and Luetchford 2012).

Throughout this period, the authors developed the idea of a comparative study of alternative food systems in different locations. The hope was to extend anthropology into a new area, and to use anthropological methods and approaches to contribute to existing debates involving academics, mainly sociologists and geographers, as well as in the mainstream press. The work would document parallels between the values and goals of participants in different places and generate an analysis around that; but at the same time, the study would explore how variations between locations and their cultures played out in particular settings. The analysis in Chapters 1, 8, 9 and 10 are the result of discussions between Pete Luetchford and Jeff Pratt on the above themes. They were drafted and redrafted by the two main authors over a period of more than a year, although the conversations on which they are based go back much further than that.

This more theoretical work was entwined in a second strand of collaboration, which was empirical rather than analytical, and involved case studies. In 2004 the authors travelled to Andalusia, Spain, on a pilot project to look at alternative food provision there. On that occasion they visited organic farms with members of the consumer cooperative La Ortiga of Seville. The first of many visits, it culminated in a longer period of data collection made possible by the generous support of the Economic and Social Research Council (ESRC). Between April and August 2008, Pete Luetchford undertook field research in Cadiz province, aided by Jeff Pratt during shorter visits to the field.

After working on the Spanish case, Jeff Pratt revisited his farming contacts in Tuscany with specific questions about the creation of new markets through direct sales and the impact of rural tourism schemes, as well as the vexed question of organics and place-of-origin certification. Completing the plan required further cases and anthropologists to carry out field research.

From the beginning it was decided that including a French study was essential. It would complement the Mediterranean cases from Italy and Spain, and, more importantly, France seemed central to key debates, given the vibrant politics behind small farmers and their contribution to local food cultures, as symbolised by the activities of the Confédération Paysanne. In 2009, Myriem Naji, a trained anthropologist, joined the team. Myriem's doctoral research examined the lives and livelihood of female weavers and their families in the Sirwa mountain region of Morocco. This multi-sited ethnography focused on labour relations along the supply chain from the production of carpets in households to the point of sale in Marrakech. Analysing economic activities and relationships in the supply chain showed how the value of commodities is intertwined with power asymmetries shaped by gender, ethnicity and class. Being a fluent French speaker with close personal links to the Tarn region meant Myriem combined local knowledge with a keen interest in economic anthropology, and knowledge and experience of fieldwork and its methods.

The final piece of the jigsaw came together when Sara Avanzino was recruited. Sara graduated from the University of Sussex with a first class degree in Anthropology and Development Studies in 2011. She then moved back to her native Piedmont, Italy, and continues to research themes related to those in this book. She is currently involved in a regional activist group, the Coordinamento Contadino Piemontese (Piedmont's Peasant Coordination group), which works on issues of access to seeds, alternative forms of certification based on a 'participatory guarantee', and on legislative practices in favor of small-scale farming activities. She now works in a centre for environmental education near Turin, as well as with her grandparents in their allotments.

It seemed appropriate to have a non-native researcher doing the English study, as it was consistent with British researchers writing the chapters on Italy and Spain. It was felt that the inclusion of England would focus the case studies firmly in Western Europe, but at the same time provide contrast, given that the UK has a much longer history of industrial agriculture, and that the tendency towards longer supply chains means that food provision is more divorced from local contexts than in the other cases.

This brief account of the long genesis of this book shows the extent to which it is a collaborative effort that is rare in academic circles. Our ambition is that it will prove uncommon in another respect, by drawing new readers towards our minority discipline of economic anthropology Just as the protagonists in our case studies are struggling to build viable alternatives against the grain of the mainstream, so we as economic anthropologists have used the prism of food to question dominant ideas on economy and rationality. In so doing we propose a frame built on the values and passions of people in their everyday lives, their relationships, and the creative energies they put into their food politics, without underestimating the complications they face. If the reader responds to these ambitions, to the voices that clamor for an alternative, and see them in a new light, the book will have served its purpose.

Pete Luetchford and Jeff Pratt
April 2013

1

Introduction

Food is at the centre of contemporary life. It is celebrity culture, sensuous pleasure, health, environment, sociability, and it is politics. This book is particularly concerned with the last of these, and more especially with the politics of where food comes from, who produced it and what is in it. While discussions in the media and by academics tend to treat different aspects of food as discrete, we maintain that the astonishing attention to its qualities and provenance is a shared reaction to broader social processes and transformations.

The background to those broader transformations is a set of dramatic changes over the last 50 years in the way food is produced and distributed. As a result many people now access more food, more cheaply, and in greater variety than at any time in history. But the revolution in provision also meets critical objections. Social and political movements have highlighted uncounted costs behind cheap food and have attempted to steer the sector in a different direction. Drawing on anthropological methods and concepts, we explore the way food has become a focal point for action (and reflection) on contemporary economic processes. Through case studies, we document the possibilities and problems people face in constructing an alternative food system, and interrogate variations between those alternatives.

The supporters of alternative food movements see the mainstream as sucking value out of their social world. Farmers find their livelihoods squeezed by the high cost of inputs for specialist industrial agriculture, and by the pressure of global competition. They see their income siphoned off by all the middlemen between the farm gate and the consumer. Some resist these processes by practising mixed farming systems and labour-intensive methods as a way of creating a more sustainable agriculture, less dependent on industrial inputs. This is one stimulus to the switch to organics. They also develop direct sales to customers to maximise the return to their labour, while some concentrate on more specialist foodstuffs which can command higher prices. For these strategies to be successful, farmers need to ally with customers who themselves view the mainstream food chain as a negative force which sucks the life and diversity out of towns, destroying livelihoods and local food cultures.

This book focuses on the movements or initiatives these people create, seen as strategies towards some measure of economic closure, some limit to the exactions of the open economy and to ever expanding market integration. The moves to greater economic closure can take many forms, from the more organised to personal projects or commitments. Some are built around locality: support for farmers' markets over supermarkets, sometimes merging with wider attempts to generate localised economies through local currency schemes. Others are built around networks, linking producers and consumers in more or less stable relationships through cooperatives, community-supported agriculture and other forms of solidarity economics. To capture the broad range of the reactions we use the terms 'initiative' and 'movement' interchangeably. But in all cases, participants share a sense of common cause against industrialised food provision, against cost-cutting, and against the obfuscation of provenance and content at the expense of quality and knowledge about food.

The phrase 'sucking the value out' is both graphic and ambiguous, deliberately so. In part these movements are concerned with monetary value, the way corporations in the food chain destroy the livelihoods of small farmers or shopkeepers. But they are also concerned with the destruction of values in the plural, whether of the social qualities of people or the aesthetic properties of things and places. These are often thought of as outside the realm of monetary value, and threatened precisely by the rationality of market and commercial calculation. These two meanings of value, sometimes opposed, sometimes combined, run through the book, whether discussing the ambitions of farmers or the quality of food.

The central chapters present four ethnographic case studies from the UK, Italy, France and Spain. They illustrate that each attempt to create alternative food systems needs to be understood in its specific context before it can be compared with others; and secondly, that although the mainstream and alternative are usually imagined as different in kind, in practice they overlap and borrow from one another. For example, initiatives designed to strengthen the independence of small farmers through the turn to 'quality' foods can end up reproducing features of the mainstream. For scholars this is an analytical problem, but for farmers and for shoppers it is more of a practical problem of negotiating between worlds, needs and desires. Even those most committed to producing and consuming alternatively make recourse to the products of the conventional economy at some point.

Food has become the most prominent area in which people try to realise an alternative economy, and there are good practical reasons for this. But, and as a consequence, this is where we can best observe contradictions and difficulties in realising alternatives to the mainstream. Before sketching out the basis of these claims, we first need to say more about the politics of food.

Contesting Food

Critics of contemporary farming dispute its claims to efficiency. They argue that farming methods are not sustainable as they depend on exploitation of non-renewable resources: soils, water and fossil fuels. They take issue with social relations in production that pay workers wages below subsistence levels and farmers prices below production costs, such that they need farm subsidies or welfare to survive. They denounce the loss of knowledge about the provenance and content of food, its traceability. There have also been growing concerns about the quality of the food we eat, such as health scares generated by industrial farming systems as well as more widespread concerns about the way food processing affects our diet and nutrition (Caplan 2000; Du Puis 2000; Lawrence 2004, 2008). The furore over horsemeat in processed food dominating the headlines as we write is but the latest in a long line of food scandals (Taylor et al. 2013). The growing power of multinational corporations over all stages of our food supply chains is the main focus of dissent on these issues.

In place of mainstream food provision, critics advocate a return to more localised, smaller scale, mixed farming systems. As we shall see in Chapter 2, many studies argue that small farmers are crucial to strategies to create a more sustainable and equitable food system. There is also overwhelming evidence that over the last 50 years these farmers have struggled to live with intense competition from large scale specialist agriculture operating in increasingly integrated global markets. If we want to understand how this plays out and what alternatives are possible, we need to explore the choices made by actual farmers in specific circumstances. We need to see how and why small farmers go under, survive or invent new pathways; we need to look at the context in which they live, their rural society, its geographical possibilities, its links with townspeople, and prevailing food cultures. How does the enveloping commercial logic of industrial farming affect the choices made by those attempting to create a different kind of food economy while also maintaining a livelihood? Crucially, as we shall see, all these aspects have a history which shapes how farmers react to the present.

All the energy and inventiveness of farming families will only have a positive outcome if others buy their food from them. What do these customers want from such food? This is a notoriously difficult question to answer, even within one setting – small-town England – before we bring in the food cultures of Mediterranean Europe. Sometimes food is considered as fuel, sometimes as a carefully balanced mix of nutrients to replenish the body. It may also be the core element in family life, conviviality and identity formation. The meal itself then becomes the focus of memories and stories about the shared experience of those around the table. These variations mean that food is imbued with social and cultural significance, and that we need to think about these issues in a broad context. Yes, there is individual taste and choice in shopping, but there are also more collective processes at work. Some are economic, determining what kind of food is available and at what price, shaping our tastes (how many ways are there of sweetening breakfast cereals?) and what constitutes value for money. Other processes are social and cultural since so many identities, distinctions and meanings are bound up with food and its consumption.

Food is only one part of a world of increasingly complex and dispersed manufacturing chains. Most of us do not know where our shoes or mobile phones were made, let alone details of people and circumstances in distant factories. Even if we did know these details, there is a limit to how much we could carry that consciousness around in our daily, semi-automatic, engagement with the things that surround us. At an abstract level, the getting and spending of money can be analysed as a complex exchange of labour. In practice this tends to be blanked out: once you have paid your money, the thing is incorporated into your life and its origins have little relevance. The same is true for much of our food; in that apt phrase, there is so much that is 'not on the label' (Lawrence 2004). But the success of Felicity Lawrence's book shows that when it comes to food there is growing dissatisfaction with this state of affairs. In many circumstances, people do want to know more about where, how and by whom their food is produced. There are many overlapping reasons for this, from personal concerns about health and nutrition to the search for specialist quality food, from a commitment to environmental action to support for a local economy.

A number of recent books have engaged politically with modern food systems. They have revealed exploitation, waste and hidden dimensions to contemporary manufacturing and retail practices (Lawrence 2004; Stuart 2009). The extraordinary disparities in food supply and the uncounted external costs of industrial production have been repeatedly exposed, both at a global level and in the relation between cities and rural hinterlands (Roberts 2008; Steel 2008). Other books cover similar ground, but then offer solutions. Michael Pollan (2006), for example, gives an account of the iniquities of contemporary food production, and concludes by seeking answers by getting closer to the source of what he eats. Colin Tudge (2007) castigates industrial and commercial practices, and sets out basic principles of nutrition and agriculture as a process based in nature; while Raj Patel (2007) sheds light on corporate power in the global economy, and sketches a political agenda for social justice in food production.

The issues raised by these books are also analysed in more scholarly work, mostly by geographers and sociologists (e.g. Goodman et al. 2012; Kneafsey et al. 2008; Morgan et al. 2006). Three main approaches dominate these writings, though there are overlaps between them: the political economy of commodity chains, actor network theory, and convention theory. We now describe the broad parameters of these debates as a prelude to our own approach.

Political economy is concerned with the way that food production is subsumed into capitalist relations and industrial processes, and the profound consequences of this for social relations and relationships with nature (Goodman & Redclift 1991). This underpins discussions of how alternatives are 'conventionalised' as they are adopted by powerful interests, and de-politicised into technical and rational problems (Guthman 2004a, 2004b). Hence, standards for organics are watered down to minimum technical criteria, or calls for food sovereignty become administrative problems about food security (Goodman et al. 2012). Political economy is precisely helpful in unpacking how powerful interests take control of the food industry, even those parts that at first sight appear alternative. We have written about this elsewhere (Pratt 2009; Luetchford & Pratt 2011) and it is an influence throughout the book, but we should also note that political economy tends to focus on the powerful and what they do. As capitalist relations, practices and commodity chains become the norm, there is the danger that everything else is either subsumed within them or become invisible. One response to this problem has been the development of actor network theory, which questions the relentless, seamless logic of modernity and capitalism, and suggests instead that social processes are contingent and contested by multiple human and non-human actors.

Political economy and actor network theory are more concerned with what people do than why they do it. They assume people have specific material interests and document how they succeed or fail in achieving their goals. Both approaches seem to be predicated on a particular rationality that looks remarkably like a Western, modern, capitalist view of the world in which people struggle over economic value. To deal with that, some analysts have turned to convention theory to consider why social and economic arrangements take a given form. Convention theorists maintain that there has to be agreement between participants about what social and economic arrangements should look like. These agreements include a diverse range of possible conventions, such as a concern with market performance, industrial efficiency, civic worth, public knowledge and environmental concern, all of which combine to build 'common perceptions of the structural context' (Morgan et al. 2006: 19). This is useful inasmuch as it identifies values attached to economic arrangements, ideas about sustainability and health for example, and how these come to be a common concern. On the other hand, we need to remain alive to the danger that identifying and documenting shared meanings presumes a bounded, harmonious society, something political economy precisely warns against.

This brief summary of academic approaches to alternative food provision allows us to identify key themes. Chief among these is the question of values, introduced earlier. In academic literature, the question of values is most often broached by suggesting that consumers are either being 'reflexive' in their shopping by taking into consideration the conditions under which things are produced (Goodman et al. 2012), or are expressing 'care' for significant others through the exercise of choice (Kneafsey et al. 2008). These kinds of generalisation tell us something about what consumers say they want to do, but they reveal little about the cultural context for their values or the different agendas people live by. In contrast, our anthropological approach through case studies documents the values people adhere to in their everyday lives, the practices they engage in to prioritise different agendas, the projects they develop, and the way contradictory commitments can pull them in different directions. The evidence is that people are adept at living with contradictions, though that often means compromise and reassessment of political and economic ambitions. The case studies illustrate these tensions, particularly those between monetary value, associated with market relations, and the search for other kinds of value in the production or consumption of food. The next task is to clarify our approach by looking at the way these themes are explored in economic anthropology.

Concepts and Oppositions in Economic Anthropology

Two premises might be said to underpin economic anthropology. First, through engagement with other societies and cultures, anthropologists have questioned the universal applicability of Western models of economic behaviour. In their research they have encountered worlds in which the most important aspects of economy involved such things as the mass slaughter of pigs or strengthening social ties by giving away yams, and this has led them to identify principles other than those predicated upon the rationality of individual actors seeking their own personal satisfaction. Second, as they have increasingly moved to study capitalist societies and cultures, anthropologists note that even here market rationality is not the only principle that underpins economic activities, despite its practical and moral force. Let us look at these two observations in more detail.

In discussing the first premise, scholars have turned repeatedly to the work of the Hungarian-born polymath Karl Polanyi (1957, 2002;

see also Dale 2010; Hann & Hart 2009) to critique Western models of economic rationality. From Polanyi, anthropologists have taken a series of oppositions, which are either central to his work or have been developed from it. The first and arguably most important of these is the distinction between market and society itself. Polanyi, it has been said, 'made the modern history of the relationship between market and society his special concern' (Hann & Hart 2009: 1). In his observations on that relationship, Polanyi was much troubled by outcomes of human suffering from attempts to build society on unregulated markets. He argued that allocating goods and services through market mechanisms created disastrous levels of inequality, necessitating a political response, a second or 'double movement', which reintroduced the social and political regulation of markets and measures for redistribution by the state. He documented this process in relation to Western Europe in the nineteenth and twentieth centuries.

From Polanyi's historical observations on market and society we get further related concepts and oppositions. One of the best known, though not one much developed by Polanyi himself (Beckert 2009: 41), is between embedded and disembedded economies. These terms are best described as the tendency for agencies (business people and corporations) to avoid social regulation and so 'disembed' themselves from societies, and the opposing tendency, to make the economy socially accountable and seek to embed it back into society and institutions. The struggle over regulation constitutes the 'double movement'. Embedding and disembedding describe contested and incomplete social and political processes, a feature they share with our own use of terms that describe movements towards an opening up and a countervailing tendency towards a closing down of the economy.

The ambitions and practices of the local food movements described in this study can be seen as examples of the move to limit or regulate the dynamic of unfettered markets. These local ambitions and strategies to regulate are understood as necessary 'because commerce knows no bounds – all markets are in a sense world markets – and this threatens local systems of control' (Hann & Hart 2009: 2). In that sense, local food movements are enrolled in longer political processes, and represent a wider or more general ambition. What we know as the market today has been constructed through a protracted, fractured, incomplete and unresolved historical struggle between the desire to escape regulation and a countervailing tendency towards social control (Carrier 1995; Lie 1991). It is precisely more local systems that the protagonists n our case studies are trying to put in place.

Polanyi's influence in economic anthropology is pervasive, though not always made explicit. It informs James Carrier's recent suggestion that ethical consumption can best be defined as a political commentary on the relation between economy and society (Carrier 2012: 3) and Stephen Gudeman's writings on 'economy's tensions' (Gudeman 2008). In both these cases, the world is understood to be made up of separate realms. On the one hand there is the market economy and its attendant logic, on the other there is society and community. These different realms are then understood in the Western model to be occupied by different kinds of people – individuals and social beings respectively – and different modes of thought, characterised as self-interest and mutuality. This distinction then draws on another of Polanyi's oppositions: that between formal means–end rationality as the basis of the market economy, and the actual (or substantive) social activities of people expressed through material relations and processes (Polanyi 1957).

The distinction between market and society runs through a very influential strand of economic anthropology, and feeds into the analysis of important economic processes. However, the distinction needs to be handled with care and certain assumptions and reservations made clear. Markets maybe outside a particular social realm but they are not outside society, and the point needs to be made because so often we hear phrases about globalisation which imply that markets just happen, as though subject to some law of nature. Instead, markets are constructed through human agency, for the transfer of particular commodities (slaves, land, coffee futures) in particular places (Athens, New Guinea, Chicago). They are also sometimes shut down. They are constructed by particular groups, or nation-states, as part of wider economic strategies. They are underpinned by specific conceptions of property and rights, which in the contemporary world are often dependent or state legislation and international agreements. We shall see that this is true for the market in organic foodstuffs and those with place-of-origin certification. Even these minor examples show that we need to look at the social (and political) processes which construct markets in order to understand what benefits flow from them, in the short term and the long term.

Finally, we need to remember that markets are not driven by animal instincts, or some pseudo-psychological cliché about human nature or rational calculus. Instead, when looking at economic activity in different contexts, including markets, we need to look critically at the concepts of the individual, self-interest and the person which are embedded in them. We shall come back to this theme after looking at the second premise underlying economic anthropology.

That second theme relates to the dominance of market rationality. For anthropologists, the study of economic life covers all the ways in which goods and services are produced, exchanged and consumed, not just those which involve money and can be quantified. This means that other principles and priorities apart from maximising individual self-interest can be found to underpin economic activity, including those exercised through such things as generosity, sharing and pooling, following the principle of communism: 'from each according to their abilities, to each according to their needs' (Graeber 2011: 95). A related idea, one that at once thickens but also complicates the soup, is that objects exchanged forge and maintain social relations. This means economic activity can also be identified with other arenas or spheres apart from markets, like households. The fact that means–end rationality is insufficient to explain all economic principles and priorities, coupled with the identification of economy in domains that lie outside of markets, has obvious implications for our study of food and alternative systems of provision. This is so because food is an important medium for forging relations that are imagined to lie outside markets, such as within families, and because the forms through which alternatives are realised, such as vegetable box schemes and local food markets, are represented first and foremost as social relations rather than ones deriving from formal market rationality. That is, the moral content of alternative markets draws on non-market idioms and ideas.

The origin of the deceptively simple observation that the economy is about people exchanging things and so enabling and perpetuating their relationships can be traced back to the French anthropologist Marcel Mauss and his 1925 essay on the gift (Mauss 2002). Gifts are, after all, quintessentially social exchanges. What, he asks, drives reciprocity, the principle that a gift given must be accepted and returned? The answer he provides, deriving from Maori culture, is that an object contains something of the spirit (*hau*) of the giver, and that this compels the recipient to return some other object.

The nature of the relation forged through gifting has been subject to much scrutiny and debate over the years (e.g. Carrier 1995; Godelier 1998; Gregory 1982; Parry 1986; Strathern 1988; Weiner 1992). There are two main reasons for this. One is that the moral and economic significance of gift-giving varies in different economic regimes. In the classic 'pre-capitalist' societies which Mauss and other anthropologists analysed, gift-giving between individuals and collectivities formed systems of exchange which frequently involved competition, power and status. For us, it tends to connote generosity and altruism, things given without thought of return, typical of social relationships characterised by pooling, helping and sharing. This is because with the ideological separation between market and society, gift-giving exists as a sphere of exchange that contrasts with the depersonalised self-interest of the market.

The second reason for debate arises because of the difficulty of distinguishing between gifts and commodities if we treat them simply as different kinds of object. The idea that gifts are personal and inalienable (in that they cannot be separated from the identity of the giver), while commodities are impersonal and alienable, is a distinction that is deeply entrenched but difficult to sustain. Gifts are often bought and commodities are frequently given away. Money itself can be given, though attitudes towards the appropriateness of money as a gift varies dramatically in different societies. One answer to this problem has been to move from categorising objects as either gifts or commodities to investigating the biographies of things as they move through social scenarios (Kopytoff 1936), for example, from being commodities to gifts to heirlooms. As Frow remarks: 'There is nothing inherent in objects that designates them as gifts; objects can almost always follow varying trajectories. Gifts are precisely not objects at all, but transactions and social relations' (Frow 1997: 124).

There is a clue here to the ambivalent status of terms like gift and commodity, which mirrors that of the concepts market and society. As nouns they seem automatically to suggest immutability, or a fixed checklist of characteristics that stand in opposition to one another. The move away from fixed and essentialised terms of reference and descriptions towards activities and processes was noted some 30 years ago (Firth 1975; Ortner 1984). More recently, it is a key feature of David Graeber's reappraisal of the concept of value. Graeber argues against the dominant Western tradition of a world made up of fixed objects and societies. Instead he says we need 'to see objects as processes, as defined by their potentials, and society as constituted primarily by actions' (Graeber 2001: 52).

These concepts from economic anthropology – market and society, gifts and commodities – are paired as oppositions and contain a highly compressed set of propositions about different kinds of economic activity, and the different morals and motivations intrinsic to economic relationships. The differences they point to are also familiar to us in our everyday lives. For example, we think of buying and selling as governed by different principles from the exchanges which take place within households or when socialising with friends. In those contexts, the significance of each kind of exchange and domain gains its force from the contrast with its opposite. They entail each other. Markets and commodities connote economic relationships based on self-interest; they are exchanges conducted in terms of monetary values by individuals whose social identities are (generally) irrelevant to the transaction. Society (and more especially community) and gifts connote economic relationships based on shared interests or altruism, conducted by socially defined persons who acknowledge their inter-dependence (Gudeman 2001). This is all useful and revealing, but we also need to be careful in our use of these oppositions, since there is a great deal of interaction between these different domains (Hinrichs 2000). Marketplace relations may not be that impersonal; people often calculate what to spend on a gift. Good concepts help us capture and describe the mutability of the social and economic landscape, and the way this is negotiated by people in their everyday interactions.

However, there is one more point to make about these concepts. People themselves also talk about them in a pure or ideal-typical way, utopias or dystopias where all goods and services are distributed by market forces, or conversely by communist principles of unaccounted sharing. The 'pure' market transaction characterised by rational calculation of personal satisfaction is opposed to the perfect gift understood as one given out of selfless love, but both positions are caricatures. Some writers have referred to concepts used in this way as metaphors, in that they encapsulate core ideas that we live by (Carrier & Wilk 2012; Gudeman 1986; Lakoff & Johnson 1980). This is an important point in relation to our usage of the terms open and closed economies.

Open and Closed in the Analysis of Alternative Food

The concepts and debates generated by the work of Mauss and Polanyi are still fundamental in economic anthropology and have shaped our approach in this book. Nevertheless, despite the substantial overlap, we have chosen to use a different frame for our arguments, that is, open and closed economies. There are a number of reasons for our preference. For a start, open and closed are part of the vocabulary used by our informants to describe their practices and ambitions, whether that is creating a closed farming system for organic agriculture or a local food economy. Sometimes they are used to describe a very empirical process of what goes into and out of the farm gate; sometimes they provide images for the kind of society they want to avoid or create.

Secondly, the terms reflect a set of analytical tensions around value, labour, production and exchange. At the core of the strategies to create relative closure in economic activity is the question of value and who enjoys the benefits from its creation – whether this is monetary value or other values. This means that we want to draw attention to production and consumption strategies, and not just to the sphere of exchange which has dominated the kind of anthropology summarised previously.

How do you grow and sell food without being dependent on the wider economic circuits of the mainstream? How do you make meals which valorise your own labour? This does not mean that we think that the exchange sphere is of secondary importance. Our informants were greatly concerned with the ways markets are organised, with how to eliminate intermediaries between producers and consumers, and about how prices can be set so as to represent a 'fair return'. The advantage of these terms is that it allows us to explore the many ways in which production, exchange and consumption are linked together in the economic strategies of those movements that seek to create alternatives to mainstream food provision.

Finally, the terms allow us to bring politics back in. Instead of abstract discussions of competition and markets we can see that the movements are struggles against the power of various forces in the food chain, including corporations which create monopolies and eliminate competition. They are, in a variety of ways, forms of resistance. Their scope varies immensely, from radical forms of anti-capitalism to campaigns around food quality. The politics are complex, whether in terms of the forms of solidarity created, the relationship with other forms of anti-capitalist struggle, or their effectiveness in their stated aims. The issues are fascinating, and the answers are bound to be multiple and disputed.

In developing our analysis of local food we have kept in mind that concepts – including open and closed – will be both ideal types (with a pure or 'metaphorical' content) and also illuminate the mutability and complexity of everyday actions. In an earlier work we pointed out that alternative ethical perspectives and practices seem to coalesce around four related strands or ideas:

> (1) *Social relations* are opposed to impersonality; (2) ideas about *boundedness* and *autonomy* reject open markets and the separation of production and consumption into distinct domains; (3) *fair prices* based around livelihoods contest intermediaries and profit, which is also a conversation about where and how value is created and how it circulates; and (4) *regulation* stands against unregulated markets. (de Neve, Luetchford & Pratt 2008: 2, original emphasis)

In these four areas, we noted that the alternative is a mirror image of the market, but that there is always a dialogue between the polarities. In this study we take the four different strands identified above and deal with them as a single pair of conceptual terms, glossed as open and closed economies.

In its 'pure' form, everything in an open economy is organised through market relations. The market is unfettered (unregulated) and unbounded, competition opens up all parts of the world and integrates them into one global market. Further, all goods and services, in all domains of life, become commodities – computers, water supplies or healthcare – and are distributed according to market forces. The agents behind this are individuals trying to maximise their material advantage;

their choices in the market are driven entirely by price. Origins are not important. The producer does not care where components come from or who buys the commodities, and the consumer has no knowledge of the production process. The connection between production and consumption is mediated by retailers, money and increasing physical and social distance. We can see this pure form of the open economy as an economic model, a political ideology, and also a 'metaphor' in the sense that it orients action.

A closed economy takes a similar kind of 'pure' form in the notion of autarky. A household or a community produces everything it needs – food, clothing, shelter, energy – through activity based on cooperation and sharing. There is no distinction between producers and consumers, no exchanges with those outside the group. This is also a model of economic activity, one which can never be completely realised, but can orient action. Autarky signals the possibility of life outside the circuits of capital, and as such, in Europe and elsewhere, conjures images of and nostalgia for a peasant or homesteader past; a kind of anti-modernity and anti-materialism that prioritises the rural over the urban and production for consumption over production for the market. It is frequently coloured by a general romantic discourse about tradition, cultural difference and rooted lives. These cultural images lie alongside the political. For example, autarky can also signal an escape from the appropriation of labour and surplus value by others. In that sense it is a metaphor for the fruits of labour remaining the personal property of the worker rather than impersonal value extracted through alienated labour under capitalism.

These ideal types of open and closed economies may orient action, but empirically they do not exist in such a pure form. Instead, as we have noted, they are relative terms, and the economic life of any individual or group will involve a variable mixture of both, from growing your own vegetables to the weekly trip to the supermarket. In addition, there are various kinds of interaction between the two economic circuits: the more dominant open system shapes the space where closure is possible, while corporations constantly try to appropriate the values of more closed worlds. For example, the major supermarkets, with their global supply chains, regularly tap into images drawn from a more closed economy, with its personalised imagery, loyalty cards and cakes just like those made by grandmother long ago. Through their labels they offer us a shopping experience which is a mix of price consciousness and caring. The use of open and closed as relative terms is designed not to blur differences but to clarify what is going on in these interactions.

This book is concerned not with the strategies of supermarkets but with people who are trying to create some measure of closure in economic circuits involving food. They frequently describe this in terms of autonomy, a more flexible and less absolute ambition than autarky. They want to achieve some degree of control over how food is produced and exchanged. Farmers do this by reducing their dependence on the inputs of industrial farming, avoiding bank loans or contractual arrangements with large-scale suppliers or buyers. Instead, they develop a range of economic strategies to achieve more closed economic circuits: integrated farming systems, the exchange of labour, pooling resources, micro-credit schemes. They form alliances with consumers in order to cut out profit-taking retailers through community supported agriculture, box schemes, farmers' markets and other kinds of direct sales. This more direct link between producers and consumers, and the knowledge that flows through it, carries over into the sphere of consumption, as we shall see in Chapter 3.

These strategies to increase autonomy involve some market relations, unlike the pure form of autarky. Farmers have to sell their goods in order to make a livelihood and purchase the other things that they need; but, if things are to be exchanged for money, how can you prevent the whole system being opened up? Part of the answer lies in trying to create markets which are more closed than those of the global chains, and by doing so keep value circulating within a particular group. That leaves the question of prices, which was one of the most interesting and discussed issues in our research. Should food prices reflect those in the open economy, where supermarkets normally provide the key point of comparison? Or should they be set by some other criteria than global competition? Surprisingly often we found an older moral notion of the 'just price': that prices should be set so that at the end of the exchange the producer can purchase something of comparable utility, hence no value is lost by either seller or buyer and no exploitation takes place. We look later at the various attempts to achieve this kind of exchange and the questions of value which arise.

A key question is how these initiatives play out, what they share and how they differ. The opening up of economies is a common experience of contemporary life, so where this is experienced as loss there is a shared concern to put in place an opposing 'double movement' that seeks to close down economic circuits. As responses to the rise of industrial foods, people tend to refer back to experiences prior to major changes in food provision when economies were more localised. This generates a set of common ideas on what to celebrate and what to contest in food practices. In our case studies we find a common rejection of 'junk food' that has unknown and possibly unhealthy content, and out-of-season fruit and vegetables, the tomatoes and strawberries one can buy all year round, but always the same standardised bland varieties. There is a clue here to a general objection to opening up economies, for such foods rebuke sociality; food production, exchange and consumption should instead be vehicles to forge relationships to people and places. Food is exceptional in this regard because it is possible to carry out, or at least imagine, different aspects of provisioning, the getting, preparing and eating, taking place 'close to home' within the domestic, social economy. In that sense food is sociability. Perhaps more importantly, food locates us. Tastes and smells are known to be keys to memory, and so food reminds us of who we are, where we belong and what we have done, especially inasmuch as these sensory qualities are attributable to provenance. The people in our different locations share the idea that it is important to know where their food comes from and the conditions under which it is produced, though the actual form this takes, is imagined or realisable, varies.

The political dimension of this move towards closure has many forms and colours, but the underlying principle runs through a host of movements, from the international movement promoting small-scale agriculture, Via Campesina, with its proposals for food sovereignty, through to the campaign for farmers' markets in Britain. We also note that the movements operate at and between different scales. A great deal of cultural and political networking occurs through personal contacts and the internet. The commonest political pattern is the development of autonomous groups networked horizontally, and these generate a very different kind of politics from the labour movements which have opposed the mainstream food chain through mobilising workers on farms or in supermarkets. On the other hand, the initiatives promoting greater closure emerge in specific places and are rooted in particular historical experiences and political cultures. From this point of view, the forms alternative initiatives take in an Andalusian *pueblo* look very different from the Transition Town movement in the UK. What is more, there is no general agreement on the extent to which greater closure should or can be realised. There is a shared politics and a common agenda, but also a range of practices and priorities between and within locales.

Outline of the Book

The next two chapters of the book deal with production and consumption. These are discussed separately, despite the fact that many activists and commentators stress the need for these alternative food movements to develop shared objectives and long-term solidarity between producers and consumers. We found the relationship to be more complicated than that. The production strategies of farmers are shaped by circumstances much wider than the consumption habits of one segment of local consumers, while food cultures are not a simple mirror of local farming systems. There are shared objectives, but in some cases there are also significant tensions between farmers and consumers because they bring different things to the table.

Chapter 2 sets out a framework for understanding the strategies of small-scale farming, drawing on work in economic anthropology to develop the concepts of open and closed economies. The cumulative effect of the development of industrial agriculture, the global integration of markets and the supermarket revolution has been a dramatic decline in the number of small farmers in Europe. Those who survive have struggled to maintain their autonomy in the face of these economic forces, firstly by minimising their dependence on industrial inputs and banks, and secondly by finding new markets. Most of these strategies are attempts to create more specialised and localised economies outside the mainstream. However, some strategies, such as the production of 'quality' foodstuffs with certifying labels and a price premium, are more complicated since they bring farmers back into wider economic circuits and open the door to commercial appropriation. All these strategies raise wider questions about attitudes towards tradition and progress, and about the political orientation of alternative food movements.

Food is grown, bought, cooked and eaten, and all these activities are shaped by complex social and cultural processes. Chapter 3 draws on a rich body of comparative anthropological research to illustrate the significance of these processes in relation to consumption. It sets out how ideas about closure informs many scenarios of food consumption, but also explores perceived advantages of more open economies. In contemporary Europe, a common interest for many people is eating adequately, spending as little as possible, buying food whose provenance is unknown and consuming it alone. At other times people spend money and energy on obtaining and cooking food, emphasising origins and qualities , creating and consuming meals in social and sometimes highly ritualised settings. Food is ingested and crucial to well-being, so its provenance and qualities take precedence over economising whenever the health of the body is a concern. The same is true when meals create or mark social and cultural identities, from the family Sunday lunch to major celebrations.

The next four chapters contain case studies from different parts of Europe. The choice of case studies is partly an outcome of the background of the field researchers, all of whom have European connections. Limiting the cases to the European region allows us to construct more focused comparisons. There are many similarities in terms of history and political cultures around food, for example, the vestiges of peasant societies in Italy, France and Spain. There are also notable differences: in Spain, peasant farmers exist alongside day workers who have gained access to land; in France there are many *néo-rural* farmers who have emigrated out of cities; and in Italy, former sharecroppers have experienced a complex transition to new modes of farming. By contrast, in England we find a longer history of industrial agriculture for urban populations, and a different food culture. Including the English case also speaks to and has parallels with the USA, and possibly other countries with more industrialised food systems. In this context we consider questionable some generalisations and comparisons, such as arguments which claim that local food in Europe is synonymous with specialist or certified products traded in elite markets for profit, in contrast to a politicised emphasis on local provenance in the USA. For example, we find the assertion that in 'the European arena, "just value" is the policy mantra, rather than an "alternative" focus on "just values" embedded in a normative communitarian discourse of social justice more commonly encountered in the USA' (Goodman et al. 2012: 68) to be a reductive picture of the alternative food movements which have emerged in Europe. It fails to recognise the rich cultures of food and practices that have developed over the centuries, or the variety and radicalism of the political projects revealed in our ethnographies.

In Chapter 4, the first of the four case studies, Jeff Pratt looks at the long-term history of small farmers in southern Tuscany from share-cropping to land-reform and the present crisis. The chapter looks in detail at the practical impact of industrial methods and global market integration on their livelihoods and on farming as an occupation. It then describes the experiments they have made in order to survive, from cooperatives to rural tourism and the search for new markets. One of the most successful strategies has been a partial return to the more labour-intensive production of specialities and quality foods, sold directly to both local people and to international tourists. It is one of many discussions in the book which opens up the paradoxes of certification.

In the Tarn district of southern France, Myriem Naji finds a concentration of organic farmers, discussed in Chapter 5. Many of them are *néo-rurals,* first generation farmers who since the 1960s have taken over marginal land and developed intensive farming methods, producing a vast range of foodstuffs without using any industrial inputs. They have drawn inspiration from an earlier generation of local farmers and from transnational ecology networks. They are activists in a range of farmers' movements, including the Confédération Paysanne, and have established very successful local markets for their food. Many in this network support the French 'de-growth' movement and articulate radical critiques of industrial farming and of consumer society more generally.

In Chapter 6, Pete Luetchford analyses another group of organic farmers in the Sierra de Cadiz in southern Spain. They include both *campesinos,* small farmers who existed at the margins of the great estates, and field labourers who occupied land in the late 1970s and established their own production cooperatives. Together they created a distribution cooperative – Pueblos Blancos – selling food locally and in the towns of western Andalusia. They also articulate a strong political critique of both industrial farming and capitalism generally, and this underpins much of their commitment to organic farming. The certification of organic food, and its commercialisation, is the focus of a lively ongoing debate. This chapter brings out very clearly the values associated with autonomy both for producers and in the complex food culture of local consumers. It also illustrates some of the problems encountered in building and sustaining alternative food networks.

In Chapter 7, the final case study, Sara Avanzino looks at a heterogeneous group who sell their produce at farmers' markets in Lewes, southern England. Some are from long established farming families, some are first-generation farmers, including a significant number from other parts of Europe; most, but not all, farm organically. The study explores the complex mixture of motives found amongst these farmers, all of whom value the autonomy which their chosen lifestyle promises, but all of whom also struggle to achieve a minimal income. They all operate in the shadow of supermarket prices, and to survive they have both to sell direct and to ask for a price premium, justified by reference to organic methods or local supply routes. There is here a fascinating interaction between money and other values, one with a clear class dimension. Farmers may hope to contribute to a localised food system but the dominant supply chains make this an uphill struggle, economically and culturally.

Following the case studies, Chapter 8 provides an overview of what we can learn about food movements from these four examples, following the strategies through from production to consumption. Despite their different histories, the farmers all face similar problems in achieving autonomy outside the mainstream. There are also important differences, not least in their approach to certification and the commercial opportunities offered by quality foods. Organising distribution through direct sales is a vital way of increasing income, but success depends a great deal on the frequency of markets and above all on local food cultures: what kinds of meals people create out of directly sourced food, and what they represent socially and symbolically. Attempts to forge closer links between producers and consumers are usually part of a wider project to build a local economy, one which is embedded in a variety of political agendas. The chapter brings out the way that production strategies, distribution methods and food cultures are all linked, and argues that outcomes have to be assessed both in terms of the economic context and people's political ambitions.

In the final chapter we take key examples from all four case studies to analyse the general characteristics of open and closed systems, the values they contain and how they are embedded in particular kinds of social relations. We are especially interested in the tensions between them and people's ambivalence about the role of money. On the one hand money is treated simply as a means of exchange and a necessary part of life; on the other it is seen more negatively, as a measure of work and of 'making' which destroys all other social and cultural values. We then make a selective raid on recent work in economic anthropology, using its insights to illuminate these tensions, while insisting that the movements we have documented have to be understood as defensive strategies against dominant economic forces, and often have an explicitly political purpose. We also suggest that the tension between these realms of value is found in many other everyday contexts, and that the different ways in which this is lived and negotiated is itself very revealing.

While the next two chapters are based around production and consumption as distinct types of activity with different rationales, the case studies themselves bring these different activities together in specific locations. The last two chapters then develop an analysis of the common principles based around various ways of generating closure by focusing on the arenas in which exchanges take place and are idealised and imagined, and the technologies, that is primarily money, that facilitate open exchanges.

What is the future of the local food movement and what are the lessons in the book? The key argument is that movements seeking greater closure in food provisioning struggle in the face of the open economy. They struggle to compete on price and struggle in terms of the range of products offered. On supermarket shelves, exotic foods share space with cheap bulk goods, artisanal products are stocked alongside imported out-of-season fruit and vegetables. How can a grower selling apples in a local market on a cold, wet autumn day compete with such convenience?

There are a number of possible answers to this question. In some areas of Europe, most of the diet is still produced locally and there is more of a tradition of buying cheap food direct from producers. Here supermarkets have as yet made fewer inroads into the local economy, though they are trying. Another answer is to rely on wealthier customers or the political commitment of people prepared to pay to support local provision, but consumers are fickle and buy strategically. A third answer is to give up on trying to compete and disappear into the hills to subsist, at which point you pose no threat and corporations can ignore you. Finally, there are attempts to build regional and international alliances based on networks of activists, producers and cooperatives to generate economies of scale and source a wider range of products. In the pages of this book you will encounter all these solutions and more to the problem of building and sustaining local systems of food provision.

2

Farming and Its Values

Jeff Pratt

This book will describe the lives and ambitions of four groups of farmers, three of them from Mediterranean Europe (Tuscany, the Tarn region of southern France, Andalusia) and one from East Sussex in England. They have varied personal histories. Some were born into independent farming families, others were share-croppers or agricultural labourers who acquired their own land, while a number are first generation farmers who left the towns. What they share is the energy they put into two strategies. One is growing food which has distinctive qualities, such as local specialities or organic vegetables, so that the food is identified by the way it is produced (how, where and by whom) rather than its origins being lost in a global chain of trade and manufacture. The second is finding ways to sell directly to consumers, a crucial way of increasing the financial reward for their labour. The two strategies are linked; in fact the way the food is sold can be part of what makes it special. In all cases there has to be a match between what is produced and what customers will buy, so we need to understand both the ambitions of farmers and what their customers want from the food they eat.

These strategies have been adopted by many small farmers throughout Europe, since the vast majority face a constant and acute problem of generating a minimal livelihood from their work. The farmers who appear in this book have pursued these ambitions in particularly innovative ways, sometimes as part of a wider and more radical purpose. Many are committed to realising values which they think lie outside the commercial world of maximising incomes. These values include the old ambition to maintain a farm intact and the land 'in good heart' to pass on to the next generation. This can merge with a more contemporary concern with sustainable farming practices, maintaining biodiversity or rolling back the worst features of industrial farming methods. It always includes respect for craftsmanship, based on personal skills, experience and inventiveness.

The difficulty all these farmers face is that these commitments have to be realised in a thoroughly commercial context. Their farming choices are shaped by a fiercely competitive market for agricultural products, within which they have to create and maintain their own more restricted markets. They cannot detach themselves completely from the world of commerce; they rely on their customers buying what they produce, and even their farms have a price. What is more, the search to create specialised and quality foods, often guaranteed by systems of certification, does not create a barrier to commercial exploitation. The values such foods embody can be turned into a price premium, and can be appropriated by corporate interests in the wider food chains.

Throughout this book we focus on the situation of small farmers, their customers, and the social and cultural significance of food. In exploring this reality we come up against a theme which has long been central in economic anthropology, the coexistence of monetary and other values, and the complicated ways they interact with each other. Classically these themes have been discussed in relation to pre-capitalist societies and their transformation by the growth of a monetised economy. Our material is drawn from daily life in contemporary Europe. In Chapter 1 we referred to debates within anthropology, and will continue to draw on the concepts which have emerged, not to sketch out a theoretical history but simply to help understand a fascinating reality. Our European case studies reveal a variety of attitudes, from those whose overwhelming priority is increasing their income to those who would do almost anything to exclude the market and its rationality from their lives. Above all, the case studies reveal the shifting, inarticulate and often paradoxical attitudes we have to food, and to the many ways in which it is produced.

Who Are These People?

In English we call them farmers, a relatively simple term. In France, Italy or Spain there are a variety of terms – *paysans*, *contadini*, *campesinos* and many others – all of them full of connotations. One strand running through them is the opposition between town and country, places with contrasting ways of life and moral values. In European history the countryside has been associated with autarky, provisioning itself through its own labour and resources. Economic activity was dominated by the production of what was immediately useful, and thrift was a prime economic virtue. Towns extracted food and labour from rural populations and were also centres of trade and a money-based economy. But money itself was morally ambivalent: it facilitated exchange and the widening of economic circuits, but it was also associated with profiteering and exploitation.

This evolving moral configuration of rural and urban economies has been cross-cut by other 'conversations' which are about progress. In the dominant (urban) version, those who live in the country are portrayed as backward, having a limited, parochial understanding of the world, 'outside society', the flow of ideas and civilisation, as one Italian city-dweller told me. They are also portrayed as conservative in their habits, resistant to change, and hence also outside material progress and modernity. Many of these connotations are covered by the English term 'peasant'. Townspeople have also generated more favourable representations of rural life, for example, as a harmonious idyll. Rural representations of urban life are less well known. In Tuscany, farmers used to refer to the town up on the hill as a place of parasites: landlords, police and priests, all living idly off the labour of the *contadini*. Andalusian field labourers had equally strong views: food was essential and those who produced it came first in the scheme of things, while everybody else came second.

For England, these themes have been explored by Raymond Williams (1973), and it would be possible to extend his analysis of town and country to cover southern Europe. Without going into the details it is worth pointing out how notions of modernity and backwardness are still current in representations of rural life, but are now contested and sometimes reversed. The farmers who figure in this book are almost all opposed to high-tech, high-energy industrial modes of farming and do not think it represents progress. They associate it with loss: of food quality, skills, livelihoods and autonomy. Above all, they do not think it is sustainable, or efficient by the criteria which matter. Many of them think that they are rediscovering the farming practices which predated the agricultural revolutions of the last 50 years. In a way they are accepting the earlier view of the countryside as static and the town as the locus of accelerating material progress and consumption. They simply reverse the moral weighting. Others do not think of themselves as simply repeating earlier experiences; their farming practices involve constant adaptation, experiment and innovation. They are not against progress, but they do think critically about the direction of travel. Either way, the term peasants is back. The activist José Bové and his colleagues have defiantly reclaimed the word *paysans* for their organisation, as we shall see in Chapter 5, on France. Even in English the term re-peasantisation has emerged to describe the reality of European farming.

What Is Their Problem?

Whatever terminology we use, statistics reveal the dramatic decline in numbers of farmers in Europe as a percentage of the total labour force.

	1930	1980	2008
France	36	8	3
Italy	47	11	2
Spain	50	14	2
UK	6	3	1

Numbers of farmers in Europe as a percentage of the labour force.
Source: Mazower (1998: 425), updated with Eurostat (2008).

The most dramatic falls are in the decades following the Second World War, but even between 2000 and 2010 Europe lost a quarter of its agricultural labour force (Eurostat 2008). These are aggregate figures for all those employed in farming, from estate managers to labourers, but within this group the hardest hit have been small farmers. They have been squeezed out of a living. There are many accounts available of how this has happened, and we need a brief summary of them because it provides an essential backdrop to the lives of the farmers we shall be describing.

There are three main processes at work, of which the first is the industrialisation of agriculture. There has been a dramatic technological revolution in farming practices, at the core of which has been the steady replacement of farm labour by industrially produced commodities: machinery, chemicals to maintain soil fertility, pesticides and herbicides to control crop loss, and more recently biotechnology to transform plant breeding. The net effect has been to control or minimise the role of natural processes and variations in farming. Each sector of production and each stage of a production cycle has been broken down as an activity, an industrial solution created and sold back to farmers as an input. This applies across the board from sowing wheat to picking olives. In most cases there are economies of scale: a combine harvester makes no sense for only 50 hectares of wheat, a milking parlour is too expensive for a dozen cows. The net result is larger units and increasing specialisation: farms, and indeed whole regions, tend to concentrate on a few crops. A further consequence is that this is a very energy intensive mode of farming, dependent on diesel oil for its machinery, natural gas and oil for its productivity. This is the central fact in arguments about the sustainability of industrial agriculture.

The stream of technological innovations has had a profound effect on farming as an occupation. What was once an activity based on land, labour and skill has become a delicate balancing act between very substantial production costs from industrial inputs and uncertain 'farm gate' returns. The result has been a substantial loss of autonomy. It has also generated a commercial logic of a kind unknown to most farmers a few generations ago. Long-term production strategies gave way to the rigour of the annual cycle, with its calculation of expenditure and returns. Farming is now treated as a business like any other, with its factors of production and its drive to maximise profits. 'We used to have agronomists to help us', one Italian farmer said to me, 'now they are all accountants'.

The second process leading to a highly competitive environment for farming is the steady integration of markets through the reduction of tariff barriers and mass transportation systems. Britain experienced this first in the nineteenth century, with its industrial population increasingly fed by wheat and meat from the Americas and from the white settler colonies. Free trade led to an exodus from farming in Britain. The same thing happened in different ways and at different times in the rest of Europe. Farmers have to compete on price with all the others in the European Union, and to a variable extent with anywhere else in the world. Fast-fattening livestock is fed on American soya; the price of cereals fluctuates with global harvests. Vegetables for northern Europe are produced using hydroponic methods under plastic in southern Spain; the price of green beans in England, as one grower lamented to us, depends on the price of labour in Kenya. This is what has driven so many out of farming, and for the survivors it is a search for comparative advantage, specialisation and lean margins.

The third process to affect them is the rapid takeover of our food supply by the supermarket chains. This has been well documented: four of them, with 900,000 employees, now sell three-quarters of the food bought in Britain, but that is not an exceptional figure (FPN 2012). The world's largest groups – Walmart, Carrefour and Tesco – are all global players. Their effect on our eating habits is part of the story in the next chapter; for farmers, supermarkets have become the crucial players in this transformation, creating the space where all the other processes take effect. It is the size of their operations, their purchasing power, which has driven down margins for farmers. It is on their shelves that the global integration of markets becomes manifest, where the green beans from Kenya undercut British production. It is their logistical operations – collection points and depots relying on bulk deliveries – which make large scale farming and specialisation so prevalent.

All these revolutions produce cheap food for the consumer: in Britain and in most of the industrial world we now spend less of our incomes on food than at any time in history, 10 per cent on average. There is considerable ambivalence about what we do and do not want to know about our food. Cheap food has many costs. Some of them are environmental: the destruction of mangrove swamps in Vietnam for prawns; the land cleared in Brazil to plant soya to keep meat prices low; the continuing saga of misused toxic pesticides. Other costs are human, since the food industry has some of the worst pay and conditions of any part of our economy: workers are poorly paid, un-unionised and rely on short-term contracts, and there are some pockets of brutal exploitation, involving, for example, migrant workers in the tomato fields of Italy and those sweating under plastic to grow lettuces in Spain. These environmental and human costs are 'external' to the corporations operating in the food chain, and not paid directly by the individual consumer, but sooner or later the bill does have to be picked up elsewhere in the economy.

The farmers in these case studies have shunned most aspects of the mainstream food system, but do their initiatives represent a realistic alternative way of feeding the world's population? The question raises complex issues, both about the actual and potential existence of food scarcity (see Lappé 2013), and arguments about sustainability, where there is a tendency to focus on narrow questions about the efficiency of any one particular technology. Critics of industrial agriculture believe that there are many ways forward. One is the elimination of food waste: between 30 and 50 per cent of edible food is discarded in fields, in manufacturing and retailing processes, and in the home (IME 2012; Stuart 2009). A second is a reduction in meat consumption, since globally livestock production is a profligate user of land, water and energy. A third is eliminating the use of arable land to produce bio-fuels. The shift towards organic farming has a place in these wider strategies, though the question of yields remains contested. Certification schemes, developed for trading purposes, create a clear-cut definition of organic farming; in practice there are a range of production techniques, some of which (like those for maintaining soil fertility) continue to be used in parts of conventional agriculture. For that reason, authors such as Colin Tudge (2007), with his concept of 'enlightened agriculture', move away from a polarised opposition between organic and conventional farming.

There is no doubt that the activists we meet in this book believe that they are, amongst other things, building a sustainable farming system for the future, but the strategies they adopt can seem ambiguous, both politically and culturally. Can you still be radical if you turn your back on a hundred years of mainstream technological innovation, market growth and consolidation? Does that mean you become a harmless pocket of nostalgia, producing expensive food for the discerning middle classes?

Tensions around this issue run through many of our conversations with farmers and their customers, and surface constantly in the research literature on alternative food chains. In our case studies, a backward-looking dimension emerges because many of the older farmers and consumers we spoke to had direct experience of a very different reality. Their farming had once been based primarily on manual labour and skill, spending little or no money on buying industrial inputs. It was a long-term endeavour, an engagement with nature more than with markets. It was a mixed farming system, with a variety of crops and livestock, because the farm had to provide the food consumed by the household, and a surplus to feed the rest of local society. In our French, Italian and Spanish examples, the vast bulk of the local diet had been locally produced in living memory. Britain has been different for longer, but even here some changes are quite recent. Up until the 1960s, a conurbation like Brighton had a rural hinterland of market gardens, dairy and livestock farms selling to greengrocers, butchers and wholesale markets (see Chapter 7). Most of this disappeared before the recent partial revival of local food production and distribution in the form of farmers' markets. Even where these 'pre-revolutionary' farming patterns are historically very remote, they leave a kind of shadow, a ghostly model of a food system as it once was and could be again. It is a model now steeped in romantic colours.

While these farmers share an opposition to industrial farming and global markets, their priorities are quite varied and do not always move in the same direction. Some are primarily concerned with the continuity of their family's livelihood, others with food quality, or environmental sustainability, or preventing commercial operators profiting from their work. We have also suggested that evoking the merits of farming practices from an earlier period may create ambiguity and doubt about the political implications of the policies advocated. In fact we can find socially radical and conservative strands coexisting in the same food movement. As outlined in Chapter 1, we deal with these variations and ambiguities by proposing an alternative framework for thinking about these initiatives around food, one which unifies their various dimensions.

Open and Closed Systems

The struggles of small farmers reveal a dynamic tension between two modes of economic activity. One aims at relative closure: to keep the fruits of labour and creativity, the goods and resources it generates, within the bounds of those who produce them. The boundaries are those of particular domains: the household, the farm, the locality. Closed systems have corresponding economic virtues and values: thrift, self-sufficiency, autonomy. Open systems are based on an expansion of trade, technological innovation, accumulation and the ever-widening circulation of goods. An entrepreneur assembles everything needed for production – capital, labour, land, materials, technology – from wherever is cheapest; thus widening the circuits potentially lowers the costs. Inevitably, the two modes tend to be represented as a contrast between tradition, with its cycles of renewal, versus the modern, with its linear progression.

No economic or social system can be completely closed; there is inevitable interaction with what lies outside it: the market. Farmers operating in a part of the economy with very rapid technological innovation and increasingly global markets face a particular problem. Most of those we spoke to believe, in the simple words of Colin Tudge, that 'agriculture really is different' and cannot be practised sustainably using the dominant business model (Tudge 2007: 177). They put their work and energy into particular values which are embedded in the relatively closed networks they have created: the long-term productivity of the farm, quality food, craft skills. They see these as threatened by the open economy, yet they must also realise some cash income from selling goods. It is this moment, 'the point of sale', which generates the most tension. Firstly, because it is the moment when work and values become fixed in a price, two worlds brought together in ways which often seem to farmers very arbitrary. Secondly, this move opens the road to possible commercial appropriation; it makes it possible (though not inevitable) that others will profit from their labours.

In order to take this further, we need to go back to anthropology, which has long insisted on a very broad definition of what constitutes the economy, covering all the ways that goods and services are produced, distributed and consumed. Only part of this activity takes place through the medium of money, or is organised through a market, though this tends to be the only part studied by economists. Since their early work in Africa and New Guinea on spheres of exchange, anthropologists have been familiar with situations where people engage in different activities within different sets of social relations, each of which has its own patterns of reciprocity, values and moral qualities. They have studied the ways in which monetised relations have spread through societies, and debated whether this brings with it a distinctive kind of morality or rationality.

For the themes covered in this book, one of the best anthropological starting places is Gudeman and Rivera's book *Conversations in Colombia* (Gudeman & Rivera 1990). Their analysis is based on a distinction between two institutions, or groupings: the house and the corporation. The house is 'locally based and wholly or partly produces its own means of maintenance; it produces as outputs some of the inputs it requires. The house is never fully engaged in or dependent on the market'. By contrast, 'The corporation is enmeshed in exchange; it buys to sell in order to make a profit... Normally, it does not reproduce its own material necessities' (ibid.: 10). The house is grounded in a locality and produces through engagement with nature; it is not static, but not geared to expansion. By contrast, the corporation is less rooted, buying in what it needs, selling its output in wider economic circuits, oriented toward profit and expansion. Neither institution constitutes a total economic system, and in later work Gudeman (2008) develops more complex accounts of the multiple domains or levels at which the global economy moves and interacts.

In our case studies, 'the house' corresponds to a farming household, with productive resources of land, livestock and work implements. Over time, both the people and the base (Gudeman and Rivera's term for these resources) have to be renewed and replaced. This means that everything the farm produces has to be allocated to different purposes – for example, some animals are consumed, others are used to renew the flock or herd; a harvest is divided between consumption, fodder and seed corn. This process of renewal is ideally achieved through a circular restocking, not purchase. It is always better to obtain what you need (animal feed, implements) from your own labour and the resources of the farm, or direct exchange with neighbours, than through spending money. Obviously, this applies equally strongly to what is often called domestic labour: cooking, cleaning, care of children and the infirm. This strong emphasis on closure lies at the heart of what Gudeman and Rivera refer to as 'the house model', and in local economic wisdom is developed in various sayings about what is inside and what is outside, about the significance of walls and doors.

However, the house does need to engage with the market for certain purposes. In Gudeman and Rivera's examples from rural Colombia, there are two occasions. One is to store the value of something perishable: fruit can be sold and the money used to obtain a different foodstuff which re-enters the house at a later point in the cycle. Another is for trade, to transform something which the house has produced in excess of their needs into something which they cannot produce themselves. In both cases the outcome of the market transaction is to replace the base, the resources of the house, not to achieve a material gain. Two related issues arise out of this situation: one is the principles which underlie the interaction between house and market, the second is the question of value.

A marketplace is any arena where a number of buyers and a number of sellers compete with each other; the outcome of that competition sets the price of the commodity in question. There are of course many factors which bear on this competition and outcome, but that is the simplest version and the result is precise and unarguable – a market price – while the mechanisms which have produced it can seem hidden or arbitrary. A farmer takes a sack of potatoes to market and she finds that there are very few buyers, or that there are many sellers, some of whom have had a bumper crop, and this reduces the return she obtains. If the purpose of going to market is to store or exchange what you have produced, the effect is compounded, because what matters is not the price of potatoes per se but that the money received will allow you to purchase a necessity of equivalent worth. What constitutes equivalence is an enduring problem, often referred to as the 'just price'. Farmers such as these do not have an intrinsic problem with exchanging goods for money when necessary, but they do have a problem with the market mechanism as a way of determining that something they have produced comes to be worth a particular sum of money. If the price is 'wrong' (does not represent equivalence) then something more than the use value of the goods has been extracted in the exchange, sucked out of the household. Gudeman and Rivera point out that this conversation about the just price has been going on since at least the time of Aristotle, and has continued amongst both economists and farming households ever since. There are not many ways of establishing equivalence; weight or volume is rarely an adequate measure, so the most common criterion is comparable labour time in production. Many farmers in our case studies hold explicitly or implicitly to some version of a labour theory of value, and believe that this is what will generate the just price.

Some farmers have no objection to prices set by market processes. Others have reservations about the mechanisms but admit that they price their goods after studying those of neighbours running market stalls or box schemes, partly because they do not have any accurate way of calculating their own production costs. Some have much stronger objections, and are wary of markets in general. They believe that part of the value of what they sell sticks in the hands of the traders and wholesalers. This gives rise to an understanding of exploitation which is rather different from that of wage labourers, and I shall return to this later. In addition, they are antagonistic to the kind of accounting systems which connote a commercial attitude towards their activity. They want a just price because that is what achieves an equitable exchange of labour. The trouble is that they struggle to find a mechanism to realise this in practice. A few try to get as close to autarky as possible, while the rest engage with markets and have to find ways to 'enclose' them, that is limit the ways in which value is sucked out of the circuits. In practice, they operate with direct-sales networks and set their prices so as to achieve a livelihood comparable to those around them. It is a roundabout way of achieving some kind of equivalence, though as we shall see their point of reference is often that of the minimum wage.

In these reactions to the market we find also a second and related issue, that of value. What are small farmers' priorities when they grow food? Clearly there are a range of objectives. At one extreme people grow food to feed themselves, including, at the margins of our research, a rapidly increasing number of people throughout Europe who grow their own in urban gardens and allotments. At the other extreme are commercial growers of organic vegetables, for example, whose priorities are those of any other business seeking to increase profits. As Gudeman and Rivera note, this is essentially the same distinction as Aristotle made between the domestic economy, where things were valued for their use, and the money economy of retail trade, where things were exchanged for gain. They discuss the way this opposition remained important in understandings of how the economy operated down to the time of Marx, who formulated two ways in which goods circulated. One was a peasant mode, or a 'natural economy' (similar to what we are calling a closed system), where commodities could be turned into money in order to reconvert them back into commodities. The second is that of the capitalist, who invests money (capital) in a commodity in order to convert it into increased money, a version of what we are calling an open system. At the same time, Marx took the concepts of use value and exchange value and reformulated them as properties of all commodities, in a world increasingly dominated by the circuits of capitalism (Gudeman & Rivera 1990: 49).

These distinctive modes of economic activity have been taken up in more recent analyses, but there are two points to make before moving forward. Firstly, the struggles for autonomy of small farmers are reactions to revolutions in the food chain: the spread of industrial farming, the global integration of markets and the corporate power of supermarkets. In this context, trying to gain autonomy implies farmers 'distancing themselves from the economic intentions of others' (ibid.: 52); in other words, some kind of closure. It extends to the realm of values, a notoriously difficult term in economic anthropology, as in everyday life. The farmers we met talk frequently about the difference between those who produce food to eat and those who produce food for profit, between quality and quantity, between the long term and the short term. They pursue a whole range of values, embedded in social groupings and networks, against the single-minded pursuit of profit in production and value for money in consumption.

However, life is rarely that simple. The farming realities described in this book do not fit neatly into one of these two boxes. The two ends of the spectrum are clear in our case studies: a closed economy with its drive for self-sufficiency, often with a strong anti-capitalist logic, and an open economy of commercially oriented farmers. In between there is a variety of strategies, as well as a variety of ways people discuss the distinctions that matter to them. In some interviews we hear people talk as though money and markets were destructive of all their values, when in practice they are only opposed to some uses of money and some kinds of exchange. Our intention is that the concepts of open and closed systems are used as relative terms which illuminate the different objectives people have and the tensions between them. It is the complex interaction between these two 'models' in the lives of producers and consumers which is so fascinating.

The New Peasantries

Gudeman and Rivera's description of the house economy as a closed system is drawn from studies of South America, though many of the details would hold for an older generation of peasants in Southern

Europe. It is based on subsistence farming, where the household produces virtually all its needs and obtains the remainder through direct exchanges or a minimal engagement with local markets. We refer to this model of maximum possible closure as 'autarky', an ambition, and a term, used by a few of the farmers who appear in this book. It also has a more diffuse presence, as a point of reference and measure for less complete forms of closure, and we drew out this strand in Chapter 1. Autarky is not a reality or ambition for most of the farmers we encountered; instead, they are thoroughly engaged with the market because they have to sell food in order to obtain cash for their other material needs. However, they still do attempt to reduce their dependence on the market through various strategies, including some which have been described for the house economy. For example, they try to ensure that as far as possible the productive base of the farm is 'self-provisioning', generating the inputs needed throughout the productive cycle rather than buying them in. This is a form of relative closure, and is often described by farmers as a struggle for autonomy.

One important analyst of this reality is Ploeg (2009). He reintroduces the term peasants into writing about contemporary agriculture, not to refer to survivors from an earlier period of history but as a model of farming which is a response to agro-industry and market integration. He claims that the number of peasants is growing, not least in Europe, and that we need more of them. The starting point for his analysis is the emergence of what he calls a food 'Empire' (ibid.: 3) which has restructured modern agriculture and controls all aspects of food provision. He describes this Empire as a regime, a series of connections between agro-industrial corporations, food processing and engineering, large retailers, state regulations, research and technology. It is quintessentially an open system, one whose profits have drained the income out of millions of farms, and in some places led to the abandonment of farming. In reaction to this Empire, Ploeg has identified a process of re-peasantisation, which is, 'in essence, a modern expression of the fight for autonomy and survival in a context of deprivation and dependency' (ibid.: 7). Ploeg's study is an ambitious synthesis, attempting to cover the dynamics of contemporary agriculture across the world, and here we can only set out a few of the arguments.

Peasants have always been on the defensive, trying to resist the extraction of landlords, traders, tax collectors and colonial officials.

Then came some version of the agro-industrial revolution, which destroyed most of their livelihoods and, so the usual account goes, history rolls over them. Ploeg argues that the history of the peasantry did not end there; they continue to exist, their lives shaped by resistance to dominant economic forces, which are now those of the new food Empire. We should stop thinking about peasants as backward and isolated from the march of progress; they are instead constantly evolving in the ways they engage with nature in order to maintain and improve a resource base and gain a livelihood. It is just that we are largely unaware of the kind of experiments and innovations they practise; they are not recognised by the dominant modes of science and technology. Ploeg has extended the normal definition of the peasantry, not least in terms of their relationship with markets, and this shines a useful light on many of the issues in this book.

An important starting point for developing greater autonomy are the 'ecological resources' of the farm; not land in the abstract, as capital, but the productivity of soils, terrain or livestock built up over the long term by creative farming practice. Using that base, farmers minimise the purchase of the inputs typical of industrial agriculture. They grow feed for their livestock, they maintain soil fertility through manure or rotations, and they extend the useful life of animals. Sometimes they exchange resources – surpluses, seeds, implements, grazing – directly with each other in what the Dutch call 'closed wallet transactions' (ibid.: 48), and what in the Tarn are called *troc* exchanges, as we shall see. In order to reduce their costs and their dependence, they create collective or cooperative organisations to purchase machinery and sell their goods. They avoid taking out loans to finance their operations, instead trying where possible to invest using money reserves accumulated from previous farming years or very commonly from income generated by off-farm employment (Ploeg 2010: 8). Last and not least, they try to increase the return on what they produce by direct marketing, rural tourism or the pursuit of quality.

We find many examples of all these strategies in our four European case studies. They are ways in which farmers distance themselves from market relations: not from localised marketplaces as such but from the food Empire of globally integrated markets. It is true that contemporary peasants, if we are to call them that, are much more engaged in buying and selling than those of previous generations.

Nevertheless, they try to minimise their economic vulnerability, achieve greater autonomy and move towards what we have called a closed system. Here there is also a challenge to earlier Marxist analysis of the impact of capitalist relations on peasantries, which generally argued that once they became dependent on wider circuits for any part of their needs the 'natural economy' had been destroyed, along with their independence. Instead the focus has moved to the degree of involvement in market relations, on ways in which people create some degree of autonomy, and on the evolving dynamics of this situation. We should also note that these strategies make the standard accounting methods used by entrepreneurs inappropriate. In these peasant strategies, we cannot divide up the farming process into factors of production, or find a 'bottom line' of profit and loss, since they are all bound up in one integrated cycle of work and renewal (Ploeg 2009: 49; Pratt 1994).

It is hard to make an overall judgement on the success of these strategies. The evidence from these studies suggests that peasant farming is more efficient than industrial farming in terms of the amount of food it can produce from a unit of land, as well as in terms of energy use and long-term sustainability. It also suggests that in some circumstances small farms can make the same income as ones twice the size, if they reduce the cost of inputs. It is also clear that industrialised and specialised farms, integrated into global markets, are not immune from failure. On the other hand, it is also true that the incomes of small farmers are very low by the standards of the societies in which they live, that they are an ageing group and that there is a continuous exodus from agriculture. Some are extraordinarily inventive in reducing their cash needs, but it is hard to do. There are other kinds of reward for this work, and other values, but there is a particular problem for an occupation which has historically been a hereditary one, where the next generation now go to school and socialise with others who are enmeshed in very different lifestyles and work routines. There is no magic formula for success. There are places where integrated farming systems are difficult to practise, where patterns of reciprocity and cooperation are hard to establish, and where there is little demand for rural tourism or direct sales of quality food. Our examples show that even in more favourable circumstances there is always a struggle.

Escape or Capture?

We have seen that there are many ways to escape the fierce competition of mainstream food chains and achieve greater autonomy: minimising the purchase of industrial inputs, establishing exchange networks and cooperatives, selling direct. If these are relatively straightforward, there is one strategy which can have unintended consequences: the attempt to increase income by selling food at higher prices because it is 'better' or 'different' from the mainstream.

This strategy builds on many features of a closed economy, but it can also potentially undermine it, and this created some of the fiercest disagreements amongst farmers in our case studies. The turn to quality builds on values intrinsic to their farming practices and the reaction to industrial agriculture. Local varieties and specialised foods, craftsmanship, the avoidance of pesticides and herbicides – all generate products which can have a niche market. However, if demand for these products grows then commercial operators can move into these areas and capture them. The key step is certification, and the establishment of regulatory bodies such as those dealing with organic food and place-of-origin certificates (for instance, the DOC system used in the wine industries of France, Italy and Spain). In each case this happens only when consolidated economic interests – local chambers of commerce and the like – invest in the project and mobilise for recognition (Barham 2003). It is only certification which enables the movement of goods out of localised and more personal networks into wider markets – the shift, in our terms, from a closed to an open system. It allows the conversion of culturally defined values into monetary value, through mechanisms explored by David Harvey in his discussion of monopoly rent (Harvey 2001; see also Pratt 2007).

There is a great deal of ambivalence about certification. For some farmers it seems an unproblematic and inevitable solution to the need to provide guarantees and reduce fraud. The bureaucracy can be tedious, but it allows their goods to circulate more widely and delivers economic benefits. Some pursue these niche markets exclusively for higher financial returns. Others, however, are much more sceptical. They set out to produce good fresh food for local people at reasonable prices and end up in a world dominated by large-scale commercial interests, in which the word 'quality', endlessly repeated in publicity material, becomes a magic passport to charging higher prices to a discerning elite. 'DOC has ruined my life', wrote the former mayor of a central Italian village, complaining about the way any taste, landscape or experience can be turned into a commodity (Veronelli and Echaurren 2003). Specifically, the vineyards of his village had been taken over by multinational companies, and his supply of cheap local wine had disappeared into expensive bottles with fancy labels.

In the chapters that follow, we shall regularly encounter arguments amongst farmers about the merits of certification, and the opposition between open and closed systems reveals one reason why this should be. In published research, the debate about food values has surfaced forcibly in many contexts, including the 'mainstreaming' of fair trade, but here we shall mention just two examples: organic farming and the Slow Food movement. Organic farming is a fascinating example of the interaction between closure and commercial circuits. In Britain, a very disparate range of social and cultural interests culminated in the formation of the Soil Association in 1946. It brought together scientists advocating the maintenance of soil fertility through the recycling of nutrients, and those who were strongly opposed to all industrial inputs into farming, sometimes opposed to industrial society altogether. Some emphasised the nutritional value of food grown organically, and some had a much more mystical and romantic concern with the health and purity of the nation (Conford 2001). Concern with health and sustainability continue to be a major impetus to farm organically, but so also are farmers' strategies for self-provisioning. Although organic certification bans chemical fertilisers and pesticides, there are approved lists of products which can be employed, and there are industries to supply them. For many in the organic movement, the ideal remains a closed farming system, and as we shall see vividly in Andalusia and France, this can be driven by a strong economic desire to stay outside capitalist circuits.

Although there are many different political strands and values in the organic movement, in order to sell food with an organic label it has to be produced according to an increasingly uniform set of regulations and monitored by a certifying body. This definition is designed to enable international trade. It has been called 'a technical fix' by Julie Guthman (2004a : 179), one that simply lists what products cannot be used in organic farming, and has been stripped of much of its original social and environmental content, to the distress of many movement activists. Her study then goes on to argue that most organic farming in California increasingly resembles the mainstream 'conventional' sector. Its fruit and vegetables are produced on large estates, using intensive methods and migrant wage labour, then trucked across the continent and mostly sold in supermarkets. These studies (starting with Buck et al. 1997) generated what has come to be known as the 'conventionalisation thesis', and a great deal of debate (e.g. Gibbon 2008; Lockie and Halpin 2005; Michelsen 2001). Nevertheless, the analysis of California holds true for much of Europe. There are large-scale operations, focused primarily on profit per hectare, which have appropriated the values which were inherent in the original movement but eviscerated most of the content. We have argued that the expansion of a commercial organic sector does not eliminate smaller producers (Luetchford & Pratt 2011; Pratt 2009), but it has created substantial differences amongst organic farmers, and has raised serious questions about who gets to define the category 'organic', and to what purpose.

A second example of this debate is the Slow Food movement, which reveals the complexity and ambivalence of building an alternative food movement around locality. Slow Food was founded in northern Italy in the late 1980s with radical ambitions to celebrate good local food and the joys of conviviality, against the snobbery of *haute cuisine* and the cultural degradation of fast food. Now it is a worldwide movement, at the heart of which are local organisations based on territory. For the founder, Petrini, territory 'has exactly the same sense as the French word *terroir*: the combination of natural factors ... and human ones (tradition and practice of cultivation) that gives a unique character to each small agricultural locality and the food grown, raised, made and cooked there' (Petrini 2001: 8). Local Slow Food organisations are built around a combination of dining clubs and initiatives for the valorisation of local speciality foods, for which the movement has developed its own system of certification.

The priorities of the Slow Food movement have evolved over the last decade, with stronger attempts being made to bring together producers and consumers in 'food communities' and reach out to the global South (Andrews 2008). At its heart remains an opposition to fast food and all the global trends in the food system summarised at the beginning of this chapter, but the proposed solutions have generated a range of criticisms (see Lotti 2010; Simonetti 2010). The most relevant issue here is the way Slow Food has commercialised the value of the local while creating the opposite of a local food system. In other words, it has generated a mechanism whereby local specialities can command a higher price, which benefits some local producers but also a whole range of commercial operators. Valorisation and certification enables the export of these local specialities, while the publicity material and Slow Food guides to shops and restaurants bring in affluent tourists. The movement once again illustrates the tension between open and closed systems, and between different conceptions of value.

The Politics of Small Farmers

Many of the farmers we shall encounter in the case studies in this book have been part of organised political parties and unions. The Sienese former share-croppers were members of the Communist Party in the reddest province in Italy, mobilised in a struggle for land reform which lasted more than ten years. Similar struggles are part of the historical legacy in the Tarn region of France, though less so amongst the new settlers of the current generation. Andalusia has a history of socialist and anarchist mobilisation amongst field labourers going back more than 150 years, and some members of the cooperative which figures in our case study took part in collective action to obtain land rights in the post-Franco period. In our British example this kind of political history is absent.

Contemporary food movements of the kind documented in this book are more difficult to place politically. Conford's study of the Soil Association brings out the extremely complex mix of political currents which have run through the organisation, from nostalgic aristocrats to anarchist doctors. All writers on the Slow Food movement devote considerable space to the analysis of its combination of popular and elitist activities. It is also difficult to generalise about the social and political orientation of customers. Are these movements providing cheap food for anybody in the locality, or expensive specialist goods for the more affluent? It is an issue which troubles the farmers. In Andalusia, their express aim is to match prices anywhere in the system, despite farming organically, but they are amazed to see the mark-up when their food is sold in the boutique shops in the cities. In Sussex, an organic farmer wanted to do the same, but admitted ruefully that he was producing 'Christmas presents'.

One problem in placing these movements is that they lie at a tangent to the kind of split between Right and Left which dominated most of Europe in the twentieth century, for example between free-marketeers and advocates of socialism or social democracy. Additionally, if we take any one strategy or indeed symbol (like the soil) of a particular movement, we will find that it can be found elsewhere, in another movement and combined with other objectives which overall lead in a very different direction. We need to look not at elements but their combination, and how they evolve in their social context. Obviously, this cannot be done in isolation from the case studies, but it is worth signalling in advance two recurring themes: the question of exploitation, and that of the politics of closure when articulated around locality.

The issue of exploitation is very important for these farmers, even if none of them are involved in workplace struggles over employment and wages. Many of these farmers' strategies are designed to prevent others making profits out of their labour as it happens not at the point of production but exchange. It is a concern for Gudeman's peasants dealing with traders, and for European farmers faced with the huge price increases between the farm gate and the shop, which they see as value extracted from them by unproductive middlemen. It is part of a struggle for autonomy, which from their perspective is often continuous with previous struggles against landlords and tax collectors.

In this sense, there is some continuity with European peasant populism, a complex movement, focused on rural poverty and exploitation, sometimes portraying the city as corrupt, sometimes evolving into 'blood and soil' nationalism. Populism today, in Europe and North America, is associated with the rights of the self-employed, independent small businesses, against the big corporations and state regulation. In the case of Italy's Northern League, these themes have been combined with the rhetoric of localism, tradition and belonging, which in turn has led them to support some of the initiatives of the Slow Food movement (Du Puis & Goodman 2005: 363). The Northern League's xenophobia is alien to the Slow Food movement, and none of the food activists who appear in this book support the more reactionary versions of populism, but these overlaps demonstrate the range of political agendas associated with appeals to autonomy, and the need to contextualise them.

A second theme is the question of 'the local' in food politics. In Mediterranean Europe, the older generation remember a time when every small town was largely fed by the farmers of the surrounding countryside. The retail revolution has blown most of this away, but the desire to recreate boundaries surfaces in all the examples. We find it in direct sales, place-of-origin certification, farmers' markets, campaigns to close out the destructive power of supermarkets, the Transition Town movement and local currency schemes. There is a complicated mix of things going on here, not least because the significance of place varies considerably in different regions.

In Andalusia, the *pueblo* is both the town and the people who inhabit it. That excluded the landlords since they did not live there. Feeding the *pueblo* combined a local and a class-based strategy, not least in anarchist practice. In Tuscany, class politics as articulated by the Left was historically in opposition to the kind of local loyalty (*campanilismo*) articulated by the centre Right in a hierarchical model of cross-class harmony. Only more recently has a more radical and democratic politics of locality started to emerge. In the south of France, political movements have espoused the small-town life of the *pays* and its autonomy from the centralising, interventionist neoliberal state (Pratt 2003: 166–70; Touraine & Dubet 1981). In Britain, meanwhile, we have communities; the word appears on every page written by the Transition Town movement (see Pinkerton and Hopkins 2009).

The concept of the local is much more conceptually nebulous and porous than it was in earlier generations because so many economic and cultural boundaries have been eroded. It has become a very powerful component of contemporary food politics, but in a way which often combines two separate themes. The local can be a celebration of distinctive traditions and knowledge, in which case it may also involve claims to authenticity or purity, both for the food and its producers. It is a short step (in one direction) to the politics of exclusion and xenophobia, a real threat in some Italian celebrations. However, support for local farming practices and foodstuffs can also be a form of economic resistance to mainstream food chains, and an essential part of the attempt to create spaces outside the global circuits of capital.

In practice, these two themes are combined in various ways. Sometimes we find networks linking producers and consumers, often people who share the same values and the same social space. Nevertheless, these networks only link up one part of the economy. The limited reach of these initiatives is partially concealed by talk of bounded localities and united communities, sometimes also conjuring up a bygone world of peasants. It is as though a (selective) antipathy to commercial values can only be articulated through reference to a bounded place (or *terroir*) and to tradition. There are also more radical initiatives, which see local food systems as just one part of a wider struggle to create a more closed economy, one set against market forces and unsustainable patterns of production and consumption.

Building a local, sustainable food system is a substantial achievement, both in itself and as a way of creating critical awareness of wider questions about economic values. However, it represents a closed system for only one part of a local economy, since farmers (and everybody else) are also customers when it comes to obtaining other goods and services. One response to this situation, pursued by some of the Tarn activists, is a strategy of maximising autarky, a dramatically reduced set of material needs which can be sourced locally. For the majority there remains the difficult question of how to 'scale-up' the operation of a closed economy so that it can handle more complex needs, divisions of labour and trade (Harvey 2012).

All this suggests that the kind of food movements explored in this book contain a complex mixture of overlapping political themes, with many tensions between them. This chapter has suggested some of the ways in which we can understand these tensions, while the central parts of the book are designed to allow each of them to be understood in its context.

3

Food and Consumption

Pete Luetchford

F ood programmes today dominate TV schedules. In the UK we are shown settings that combine production and consumption, such as idyllic *River Cottage*, while contrasting programmes follow the hustle and stress of commercial restaurants, like *Ramsey's Kitchen Nightmares*. We watch people prepare and consume food to win prizes in *Come Dine with Me*, see cook-offs in *Masterchef*, and track couples trying to run more successful businesses than their rivals in *The Restaurant*. The competitions contrast with educational programmes. We go on gastronomic tours with *The Hairy Bikers*, follow Jamie Oliver as he prepares *30-minute Meals*, learn the science of food in *Gastronuts*, and look at innovation in *The Food that Made Millions*. More combative programmes like *Hugh's Fish Fight* reveal the consequences of modern production regimes, while *Jamie's Food Revolution* and *Really Disgusting Foods* connect industrially produced foods to issues of class and health.

The proliferation of programmes, symptomatic of what one author has called the 'foodification of everything' (Poole 2011), provide a window into what people want from food, but the messages are ambiguous and contradictory. The central argument of this chapter is that food is the focus of media attention because it encapsulates and embodies key tensions and contradictions in contemporary life. For consumers, the problem can be summarised as a tension between knowing about the conditions under which food is made and a supply system that largely obscures those details. Questions then arise as to why many people want to know more about where their food comes from and who made it, what obscures provenance, and what scope there is to counter that mystery?

There is no straightforward answer to these questions because various political, economic and cultural strands pull in different directions. In this book we deal with that by identifying a broad and general tendency: the way in which the opening up of economic relations meets counter-attempts to imagine and build closure in economic relations. The opening up of the economy is impelled by retailers seeking profit, and in the process putting the squeeze both on their competitors and on farmers, and at the same time attracting consumers with cheaper prices. In that sense, the open economy is driven by self-interest and 'economising' in capitalist markets. By contrast, closure is inspired by an ethic and politics of creating social relationships, in this case through the medium of food.

At one level the politics and processes are straightforward. While the more open economy provides cheaper, affordable food that widens access to a vast range of products, critics point out that there are uncounted and hidden social and environmental costs. The obfuscation of the conditions under which food is made is answered by revealing those costs. For the mainstream food industry, this comes retrospectively in the form of traceability – buying British meat to avoid unwittingly buying horse – and governance of supply chains. From an alternative perspective, it is also about tracing the pathways to food by knowing the grower or the maker. So traceability is about establishing social, rather, or more commonly as well as, formal economic connections.

On the other hand, there are complications. Mainstream retailers also signal provenance, doing so to attract customers as well as generate profit margins on the premium for distinctive qualities. Likewise, farmers and retailers selling locally are motivated by profit and have to be business-like to survive (Grasseni 2003). The difference is observable in consumer practices, especially how they construct and negotiate the distinction between products from relatively open and more closed notions of economy. For many of the protagonists in our case studies, this is about cultural context and building a viable politics, but people also manipulate distinctions according to social and economic capital, or class, as well as more nebulous features of consumer culture, such as fantasy, desire and indulgence.

There are, then, considerable overlaps and tensions between the two strands as they draw on and borrow from one another. After outlining those tensions and the relevance of the terms open and closed to food consumption, this chapter identifies their manifestation through different 'levels' in our relationship to food, from the individual body to wider social contexts. While it is recognised that one outcome of differences between foods is the means these provide to construct and maintain distinctions, such as class hierarchies, this is not the principle motivation. Rather, as the case studies will show, the common political concern of people from different backgrounds is to counter the growing separation of consumers and consumption from producers and production, which many people, ourselves included, understand as intrinsic and inevitable in the dominant capitalist economy. In pursuing that project, it makes sense to begin with food.

Food: Tensions and Contradictions

At one level, food is perfectly suited to enterprise. The TV programmes mentioned above reproduce and mimic entrepreneurship through competitions, and cooking skills readily convert into successful business models and branding. Jamie Oliver is the fastest selling non-fiction author in history. Enterprise culture driven by an ethic of money-making and opportunity also rebounds into food shopping. A barrage of messages urge people to exercise thrift and maximise value for money, and saving money justifies luxuries (Miller 1998). Supermarkets are adept at indulging this ethic. They combine the temptations of strategically placed treats in the checkout aisles with two-for-one offers, and they engage in price wars on our behalf. Eating, too, is at one level a supremely individual act. TV dinners and feeding at the fridge door seem to encapsulate an individualistic, anomic society.

In contrast to this individual and enterprise model, food is also understood as social and cultural It is something that binds people together and transcends personal desire or profit. Sharing food and table – commensality – is an act and an occasion in which social and cultural values are lived and reproduced. It is of utmost importance socially and culturally who eats together, but also who does not. In that sense, commensality is about exclusion as much as inclusion. Another powerful set of distinctions is the kinds of food different groups eat, and the occasions on which they eat them. As Mary Douglas (1975) showed, food provides complex codes to denote socio-cultural groups. Shunning pork or meat does not just set Jews or Brahmins apart; eating kosher and being vegetarian is a means to enact membership of these groups, its cultures, its beliefs and its life cycles. In the same way, eating turkey at Christmas or Thanksgiving is important in Anglo-American culture to periodically reconstitute family membership. Here, food and consumption are beyond financial reckoning because what is emphasised is the social connections that food affords. Shared appreciation and conviviality make and remake social groupings, and are not priced. At the same time, emphasising groups or communities tends to flatten out internal differences and power relations.

Here we arrive at two related conundrums that complicate social and cultural dimensions of food. First, food consumption does not just mark groups; differences between foods underpin hierarchies, and so food differences demonstrate cultural capital. As vegetarians, Brahmins position themselves (and are positioned) as the highest caste. Of course contexts and hierarchies are differently constituted, and not just in western Europe, so the similarities and differences between the coming case studies are instructive. An important common marker, however, is money, which cements distinctions and hierarchies by indexing the ability to pay. At the same time, money allows the circulation of food qualities in open markets; certification, branding and celebrity endorsement justify premium prices, and so reproduce class distinctions.

This introduces the second conundrum, which falls in two parts. The first part concerns what is hidden from view in modern food provision, the kinds of information people receive, and the extent to which certification and branding mislead for commercial ends (Lawrence 2004). The TV programmes that critique mainstream food provision say this is problematic: food quality is undermined by market capitalism. The second part refers to how shoppers respond to advertising and other messages. There may be legal statutes regarding trade description, but the information is framed in specific ways, and shoppers engage with the things on sale with their imagination – this being part of the creative process they enjoy. Beyond formal requirements, there is no need for words and images to correspond to any kind of reality. As I have argued elsewhere, evoking small farmers misleads and obscures key elements and actors in coffee production (Luetchford 2008b, 2012a). In effect, all that needs to happen is for messages to be interpreted in a way that triggers a purchase, so that the shopper can feel they are exercising freedom to make the 'right' choice for them (Carrier & Wilk 2012).

Open and Closed Economies of Consumption

The two strands of economy I label above as entrepreneurial and social correspond to what others, not least Polanyi, call market and society. In Chapter 1 we argued the terms 'open' and 'closed' better capture these two aspects of economy, not least the fact that the extent to which they can be realised varies. The peasant house economy discussed in the previous chapter corresponded more closely to the autarkic ideal of a closed economy than present attempts to create closed spaces through what are often little more than consumer imaginaries about fair trade and organics (Varul 2008; Wright 2004). Further, in our Italian study we see how that closed peasant economy has been steadily eroded. Likewise, in the Spanish case, the local village or *pueblo* economy is slowly being broken open by supermarkets, a process that has gone even further in Britain and France. As the open economy becomes ascendant in neoliberal markets, it increasingly generates the context in which the closed economy must operate. On the other hand, the further the opening-up process goes, the stronger the potential for response, particularly for some products and for certain people. The result is a kind of dual food economy in which the two strands lie alongside one another and are intertwined in everyday life.

So, as our case studies and the coming discussion of food will show, a complex picture emerges in which degrees of openness and closure are intertwined. But before entertaining that complexity in the world, it is worth taking time to reiterate what we mean by open and closed in relation to market exchange and consumption. We consider open and closed useful for two reasons. First, they aid our analysis, and in that respect they help organise the material and ideas. Second, we think people pursue and use concepts that correspond to open and closed to orient their actions in the economy. In this respect they encompass broad sets of ideas and agendas that people adhere to or try to realise in their everyday lives.

The open economy revolves around market exchange as this is currently understood and manifest in the dominant, global, neoliberal economy. In the past, other kinds of markets were more central to everyday life, and these markets as places continue to exist on the fringes, and to a degree have been revitalised as farmers' markets. Exchanges in the open market rely on the standard measure of value – money – to facilitate commerce between people independently of the context in which they operate. Again, this does not mean there are not other more social uses to be made of money, as personal gifts, for example (Parry & Bloch 1989). However, the dominant understanding of money at present is as an impersonal medium of exchange used by people to make rational decisions about how to allocate resources. In this scenario, people and markets are formally free, in that social relationships do not count in deciding the terms of exchange. Rather, decisions are made on the basis of individual self-interest. Strong claims are made about the virtues of formal freedom, which makes it look ethical (Luetchford 2012b). The ethical status comes from the way formal freedom is thought to constitute and foster creativity in production, as entrepreneurs are given free rein to develop and promote goods and services in open competition with one another. The result is imputed efficiency, which allows the best products to be made at the lowest possible price. Consumers drive this process: as they seek to maximise value they are also creative, constituting themselves as self-made subjects through shopping decisions made as formally free agents (Foucault 1991). This is the ethos and practice of neoliberalism, and the basis of the ideology of modern consumption.

Freedom and choice are therefore key words for proponents of the open economy. Entrepreneurs are free to create and promote goods and services, and profit from their endeavours; consumers are free to choose between brands and products that compete for their attention and purses. The open economy model is clearly related to globalisation, for it is the dominant way the economy is now organised, and the exercise of choice by companies and individuals means that products and ideas are increasingly produced, distributed and consumed on a global scale, and made available and affordable to increasing numbers of people across the globe.

The second strand stands as a counterpoint to the first. If the open economy globalises, the closed economy localises. A closed economy is one in which decisions about exchange are made using social criteria rather than personal preference or gain. These social criteria regulate, circumscribe and modulate by introducing ideas such as welfare, responsibility, need, compassion and love. The closed economy seeks balance and the reproduction of relationships rather than individual gain; its purest expression in the Western cultural imagination would be the household economy described in Chapter 2. The closed economy is generally evoked as a rebuke to the open system. This locates it within a romantic reaction to the Enlightenment, modernity and neoliberal capitalism, a narrative of loss in which distinctive cultures are undermined by processes of homogenisation (Kahn 1995; Pratt 2007). The unique and authentic contrasts with the mechanically produced, the organic rejects the synthetic, quality stands in opposition to quantity, diversity to singularity, and metaphors of the timeless contradict innovation and progress. There is a gradual loss of roots and authenticity, as the world is thought to increasingly look the same everywhere. Fast food is an important focus of this critique (Wilk 2006).

Welded to the romantic narrative of loss are more hard-nosed objections to the open economy as a destructive 'race to the bottom'. In one version, the criticism highlights how competition between entrepreneurs creates downward pressure on production costs and standards. At the other end, the consumer is held to account: seeking through choices the best product or maximum calories at the lowest cost drives down prices and precipitates the race to the bottom. Advocates of the open economy say the overall benefits of the expansive economy outweigh negative outcomes; critics say that it is the nature of the open economy to consign costs to another part of the jungle.

The problem is that these models, though they seem to stand for separate things and reflect different ways of engaging with the economy, are intertwined, with the open economy setting much of the context in which practices and imaginings about closure operate. As a result, people have constantly to compromise their terms of engagement. Take the case of ethical consumption, defined as the practice of giving moral consideration to what to buy, and by doing so making a comment on the relation between economy and society (Carrier 2012). There are a number of difficulties in trying to achieve ethical or political ends in the market. First, the open economy identifies the social concerns of shoppers, and packages and sells them for profit, flagging this as corporate social responsibility. Second, shopping as political action must face the tension between more personal goals, like health, and more transcendent aims, such as a concern for distant others whom the shopper may never meet. Realising these latter kinds of goals can be complicated. How can a British shopper compare a fair-trade organic apple from Argentina with a conventional one grown locally? Different ethical ideas pull in different directions (Littler 2011). Third, the mark-up on the price of ethical products – be they Fair Trade, local or organic – means that consumers must choose between their social concerns and the desire for value for money as they shop. Economic means and social differences of class then become a notable feature of ethical consumption.

All this takes place in a context of increasingly open markets. As Michael Pollan points out, the industrial food chain mixes up and renders opaque the provenance and history of food (Pollan 2006: 115). It does this through the open economy, bringing in components procured in different places, blending and transforming them. The result is a processed mix that could contain anything and come from anywhere. To learn that a pineapple you eat is from Costa Rica tells us little about it, apart from the fact it has flown a long way. An important element in this process is the ability to preserve foods for longer and transport them more efficiently (Goody 1982). So, an outcome of the open economy is food whose content is unknown and provenance uncertain.

This has obvious consequences. As the distance between producers and consumers expands, there is an increasing potential loss of knowledge of the life-worlds of the respective parties. As a result, retailers have to create and represent products in ways that try to match what they think consumers might want, and consumers have to 'fill in' by reading what they can from limited information, using their imagination. Supermarkets are powerful agents who carve out and operate in that space between consumption and production. All retailers do this to a degree, but supermarkets now dominate food retailing.

Their control is enacted through product lines which cover all bases. For some commodities, the link between consumption and production is erased as far as possible, symbolised in the plain, often blue and white label: pile high and sell cheap, but in bulk. Any losses can in any case be recouped by cutting margins paid to producers, or subsidised by premiums on other more specialist products. For some of these specialist premium products, the brand emphasises the qualities (aromas, tastes, textures) and closed, social-consumption scenarios, like a family enjoying their breakfast cereal. Here, the consumer is left to imagine (or not) where and how the qualities were produced. For other specialist products, the knowledge is flagged and a premium is charged; images and information about producers and production systems sell. Ethical ideas about closure get built into capitalism as profit.

Ultimately, these processes are an outcome of the relation between the use of money in open economy and social values. But for now we need to take seriously the ambition to act on social and economic processes, even if those ambitions are somewhat thwarted or distorted. While much ethical consumption takes place through mainstream markets and seems compromised, we also need to recognise that ethics and politics have long been a concern in consumption, and they are increasingly visible in many areas of social life (Carrier & Luetchford 2012; de Neve, Luetchford, Pratt & Wood 2008; Lewis & Potter 2011). As our case studies show, many people constantly develop and maintain alternative economic practices. In these cases, profit-seeking is understood to be absent or a secondary consideration, whether they are employing money as a medium in exchanges or not. One question concerns what people seek politically and socially in the economy; a second concerns what they do or get.

There are good reasons food is at the forefront of efforts to imagine and create an alternative economy. Much activity around food takes place outside the money economy. Food can be produced almost for free on an allotment, the unpaid housewife does food shopping in her 'spare time' though it is hardly leisure, food preparation is unpaid, and cooking at home is idealised as a hobby rather than work. By classifying productive activities like growing food, shopping and cooking as 'not work', food fudges the modern, Western distinction between the formal world of production for profit in the economy and the non-economic sociality of the home. Food also features in a greater number and wider range of exchanges than anything else, and so penetrates all areas of our lives. Furthermore, food has both personal and social significance. It nourishes our physical selves, and we appreciate tastes and textures as an individual experience, incorporating food into our bodies as pure use value; but food also maintains kin groups and larger social polities with shared food habits and cultures. Lastly, food as a product of nature, worked upon in its preparation, has unique qualities that make it authentic. No two dishes are quite the same, although industrially produced food comes close to being mechanically reproduced, like TVs or pins, and sails close to the wind of the inauthentic. These features combine to make it difficult to put a money value on food: what price a good square meal?

Food is therefore part of the economy, but it encompasses values beyond money, and there is a common idea expressed by the protagonists in our case studies that the reduction of social and cultural values to monetary equivalence requires guarding against. In the next section we show how food practices involve closing off distinct areas of life, from the body to the nation. In this closing off there is a making of cultures and politics, as people contest the open economy to varying degrees. But the attempts are always partially compromised, so how people as both producers and consumers negotiate between the opportunities and constraints of open and closed economies becomes a key issue.

Food Consumption: Imagining a Closed Economy

Through food, an ideology of consumption focusing on economic closure is imagined and partially lived in consumer societies, such as those of western Europe. Here the focus is on consumption, though it includes what are strictly speaking productive (though unremunerated) acts, like shopping and cooking. The account encompasses the most intimate and arguably closed area of our lives, bodily processes, to a discussion of the broader frames of the nation, and national cuisines. On this journey the discussion entertains other areas of life: the family, the local, the community, the village, town, county or region. In all these cases, a closing down and closing off is sought, as people pursue relationships and practise sociality through food. On the other hand, these attempts often occur in the shadow of the dominant open economy because social values become a source for profit and tend to reinforce hierarchies such as class.

When the aim is to satisfy the individual need for calories, food as body fuel, it matters little where it comes from and who made it. There are also satisfactions enjoyed by many people as they consume processed foods; tastes attached to saturated fats melting in the mouth, the sweetness of high-fructose corn syrup, or the convenience of the TV dinner, for example. That is, modern industrial foods have attractions to many people, some or all of the time. The aim is calorific satisfaction at the cheapest price. Evidence shows that mainstream supply chains have been remarkably successful at this. As obesity soars, the proportion of household income spent on food in the UK has declined, for example, from 33 per cent in 1958 to 18 per cent in 1992, to 10 per cent today (Warde 1997: 23). Part of the attraction of cheap food is that you don't need to think about where it comes from or what it does to your body.

On the other hand, many people in the case studies in this book do want to know where their food comes from and what it contains, at least some of the time. To that end, they share a desire to close down the open economy to counteract unknown content and provenance. These efforts to counteract the industrialisation of the food system vary greatly. They are a reaction to the modern and often hidden costs of mainstream food provision, and because industrial food has multiple effects, so the response is broad and multi-stranded. Indeed one could say that where a hidden cost is identified, so there is a response against it. People worry variously, and to different degrees, about the effects of industrial food on their bodies, or on the environment, or the social fabric of where they live. But there is more to it than personal interest because there is a consensus that modern food provision generally has negative effects. So although different people in different places want different things, the conviction they share is that modern, capitalist food provision corrodes societies and cultures. Accordingly, and in keeping with a common concern to roll back the process of capitalist modernity, the alternative acknowledges the places, practices and histories behind food provision by noting special qualities imparted by provenance, and by emphasising connections between producers and consumers, or by linking production conditions to consumption scenarios. The fact that mainstream retailers borrow from this alternative and that the repercussions are often refracted through class should not blind us to political intent.

These concerns clearly take us beyond the personal, but closure begins with the individual and the body as biological. It makes sense to start here because 'whatever else food is about, it is always ingested, digested, and excreted through the act of consumption' (Fine 2002: 212). The body incorporates in food the full force of the open economy, but at the same time provides a site for resistance to it. That full force, culturally speaking, is most visible in public display. In TV franchises, overweight bodies – as in the programme *Embarrassing Bodies* – or extremely thin adolescent models are types against which we can measure our own forms. By controlling and regulating food intake, we generate 'normal' bodies, or pathologically fat or anorexic ones. Much of this is framed as a choice about diet. In a world where we can eat just about anything we like when we want, diet becomes a matter for personal responsibility, and one's body an object for self-surveillance. But outward forms are also symptomatic of bodily processes; if 'we are what we eat', and we can eat pretty much anything, then we need to know what we are eating to make the right food choices. That is the omnivore's paradox (Fischler 1980; Pollan 2006; Warde 1997: 30–32).

Angst about body shape is thus overshadowed by a deeper concern about what foods contain. In this way, obesity becomes a clinical condition linked to lifestyle choices, and the epidemic of obesity – over a quarter of adults in the UK in 2012 were technically obese, according to the NHS (2012) – is taken as evidence for the damage 'junk food', with its high levels of cheap saturated fats and sugars, can do. What is worse, most if not all food contains additives, residues and sometimes pathogens. Salmonella in eggs, BSE in cattle, cancerous Sudan I dye in sauces, dioxins in pork, e-coli in vegetables – these are all invisible risks with uncertain consequences that intensify the angst about what to eat. When diet is a choice, then avoiding such things is largely framed as personal responsibility; eat cheap food with unknown content and you are bound to get fat, or ill, or worse.

The accent on individual responsibility makes it easy to dismiss alternative food provision as a middle-class neurosis about health and body image, and these are important concerns. Here the important point is that the focus on a body ethic operates within and often reproduces terms and ideas of neoliberal individualism. While there is an element of truth that the 'personal is political' and that such changes can prefigure wider patterns, diverting concerns onto individual responsibility masks class differences and the social and political aspects of economy. But the fact is that the avoidance of risks in food is not just personal. A notable feature of the rise of organics as a social phenomenon is the focus on families and children. That is, uncertainties about mainstream food and its effects on the body and health are transposed onto social forms, in the first instance the family: as important as not poisoning oneself is not poisoning the kids. Today one struggles to buy processed baby food that is not certified organic. In Spain, organic shopkeepers noted the prevalence of mothers with young children in their clientele, as did the stallholders of the farmers' markets in our UK study. This is borne out further by statistics: in one survey, the percentage of shoppers buying organic food in the past six months peaked at 71 per cent among people aged 35 to 44, a demographic group likely to have young children (Cottingham & Winkler 2007: 38).

The main concern is to look beyond outward bodily form or the effects of food on health to the possibilities food conjures for an alternative to the dominant economy. One way to think about this is to see food as a medium through which we engage with the world, and the senses as the means to that engagement. Much has been written, after Marx, about the alienating experience of modern life as rupture, the tendency to live our lives through things we did not produce, and forms of market exchange that make an individualised life inevitable. The obverse would be a world in which the producer and consumer is the same person, or the two activities collapse into one another. The value of things then resides in their qualities rather than monetary exchange values, and those qualities are directly experienced through taste, touch, smell, sight and hearing. Nothing engages the senses like food, and in no activity do we recuperate the full use value of things so completely and directly as through eating food. The smell of home-baked bread or the taste of a tomato is a powerful trigger to the senses and to memory, which provides us with context in the world. Tastes and smells linger in our memory, and when they are evoked they remind us who we are and where we are from. Big retailers know the power of the sensory experience of food; they pump the smell of bread down the aisles or bake dough on site from frozen, and they place the most visually sensuous foods, the fruit and vegetables, at the entrance to their stores. But many people, when made aware of such tactics, would consider it a 'trick', a travesty and an appropriation of an authentic sensory experience. This is because our experience of food, the way it plays on our senses and places us in the world, is unique to each of us; any attempt to reproduce that 'artificially' and en masse threatens our sense of authenticity.

That authenticity is anchored in culture, the meanings and relations we share with other people, as these are located in places and as the outcome of specific histories. Food is a primary material in constructing relationships that are deeply cultural and meaningful. It represents, perhaps more than any other substance, shared traditions as these have developed over time in particular places. By contrast, industrial food in its most extreme form, as a kind of 'body fuel', seems to deny culture. As it is for individual satisfaction, there is no sociality, and as it comes from nowhere it negates place. It is a denial and a forgetting. While this makes industrial, processed food suited to certain kinds of practice (such as convenience, or an expression of one's modernism) it does not encompass all that people want and experience in food.

The key to moving from understanding food as individual need to food as culture and sociality is memory, as this locates people in time and place. The best-known meditation on this theme is in Marcel Proust's *Remembrance of Things Past*. When the narrator's mother persuades him to take tea and Madeleine cake, memories flood back of childhood visits to relatives, and he recalls the town, the houses, the gardens and the country roads around his aunt's village. The narrator struggles to trace the source of the memories and finds them in the lime-flower tea and cake his aunt gave him every Sunday. The passage tells us that smell and taste provide the trigger to a much broader set of memories. These remain locked up until the sensory key opens up what Proust calls 'the vast structure of recollection' (Proust 2006: 63).

The fact that our memories are coded through the senses has a number of important repercussions. First, and perhaps most obviously, food provides a particularly rich sensory palette but is destroyed in its consumption, so it is strongly evocative but it requires effort to place and fix the source of the memory, a process vividly described and analysed by Proust. Food is a great medium for remembrance because memory is so tied up with tastes and smells (Sutton 2001). By contrast, 'processed foods by their very nature would seem to work to decrease knowledge and memory' (ibid.: 142). Food as remembering and memorialising contrasts with food as obscuring and forgetting. It is notable that the cheapest foods, the everyday value brands, have the plainest packaging to aid the consumer in the process of hiding and forgetting; there is no attempt to spin stories about the sociality of production or consumption.

Reinforcing the role of food as central to memory, and providing a link to sociality, is the requirement to repeatedly and regularly consume to sustain the body. This means food structures the temporality of individual life, but it also marks the passage of social time. The family members eat breakfast together, go to work or school, and reconvene in the evening. As meals are consumed, and then repeated, they constitute the fabric of lived social relations. In some societies this making of relationships through repeated activities in particular places is taken further. The anthropologist Janet Carsten tells us that Malaysians on the island of Langkawi become kin by eating together and sharing the 'heat of the hearth' (Carsten 1997). They change the aphorism 'a family that eats together stays together' on its head; those that eat together and share substance become family.

In Western and other cultures, the cooking and eating of meals are rich occasions for reaffirming closed social relations. This is achieved through specific mechanisms and, as with issues around health and the body, 'junk' food like pizza and pot noodles provide contrast. At one level the meal is about taking the generic product from the supermarket shelf and making it matter by transforming it. That bag of carrots, indistinguishable from others snatched unthinkingly from the supermarket shelf, is reworked into its given form and takes its place among other meaningful elements in a Sunday roast. Fast food, ready meals and TV dinners require no such transformation, and seem less meaningful as a result. In the transformation of generic commodities into meaningful social material, specific roles are lived and reproduced. The clichéd image is of the housewife cooking and the man carving the roast. In the UK, informants liked to stress the importance of eating family meals as an occasion, or lamented that work schedules meant that the adults tended to eat separately from their children.

The extension of personal care and responsibility from the individual to families may appear obvious, natural or right, but we need to recognise it as symptomatic of a more general process of sociality achieved through food. This sociality operates through closed relationships. In the same way that food consumption ends up inscribed on and in the body in positive and negative ways, so food practices are written into social relations. A vivid demonstration of this is found in the work of Sobo, who tells us that in rural Jamaica kin share their wealth and feed each other, so cultural logic has it that 'people tied into a network of kin are always plump and never wealthy' (Sobo 1994: 132). By contrast, thinness is taken as proof and evidence for villagers being mean and stingy.

The idea that giving and receiving – or reciprocity – is a means to create social relations is foundational to the discipline of anthropology (Malinowski 2007; Mauss 2002) and is part of a more general process of sociality as mutual recognition (Robbins 2009). Nor should it be a surprise that shopping, as the principle avenue for getting stuff in Western society, is used as a way to express relationships, like love (Miller 1998). But let us not confuse the means with the message that people use things to forge, contest or even end relationships. Food provides particularly rich material in this endeavour, partly because in people's ethical imaginations the production, distribution and consumption of food should take place in and constitute sociality. In this respect food provides a kind of symbolic capital to resist the market, as it offers a different measure of value to money. Even when food is bought in a market transaction, as it is for the most part in contemporary economies, it is quickly reworked into socially and culturally meaningful projects.

The daily repetition of basic meals sustains and reproduces groups, but there are other cycles of food consumption. In the European cultures represented in our case studies, everyday basic reproduction contrasts with feasts, which confirm and reproduce more extended relationships. Much as holidays gather meaning because they break up normal work patterns, so feasts as special occasions contrast with everyday commensality. As the anthropologist Carole Counihan explains, 'In rural Sardinia in the past there was an ironclad ethic and practice of consumption: daily consumption took place within the family and was parsimonious; festive consumption took place within society-at-large and was prodigal' (Counihan 1984: 53). Excess in public food consumption is recorded from ancient Greek culture to accounts of medieval banquets (Strong 2003), and it continues to be central in present day events such as weddings.

The more general point is that food consumption scenarios are an indicator of intimacy and distance in social relations. The family meal is one kind of interaction, a Christmas feast held by an extended family another. Mary Douglas contrasts drinks which are for 'strangers, acquaintances, workmen and family' with meals for 'family, close friends, honoured guests' (Douglas 1975: 256). While it is tempting to understand feasts as occasions in which social groups form communal bonds and cement alliances, the expression of intimacy and distance should alert us to processes of inclusion and exclusion, or power relations. Food and power are closely intertwined. Access to food has long been culturally regulated through sumptuary law. In the medieval period in Britain, for example, laws controlled both the type of food and the quantities available to different people. Banquets were divided into high table and poorer guests: '[at] Richard III's coronation feast only the king's table had three courses; the lords and ladies had two and commoners only one. The lords and ladies got the lesser delicacies; the king alone ate Peacock' (Strong 2003: 104). Regulation of access to food was premised on social hierarchies. The contrast is with regulation in the open market in which feast and famine (and who lives and dies) are an outcome of formal economic and impersonal processes, rather than social and cultural distinctions.

The broader point is that food is an important marker for flagging similarities and differences between groups of people. At the most general level, national cuisines and religious observances provide similar opportunities to demarcate social and cultural difference. Indeed, creating a national cuisine is an important part of constructing a sense of national identity – in order for Belize to become a proper nation it needed to invent a distinct cuisine (Wilk 1995). But national cultures and national cuisines tend, like terms such as family or community, to deny differences within groups. If the open economy obscures how and where things were produced, the closed economy suggests consensus, so differences of class or gender seem to disappear. The open economy makes food affordable to many at a cost to society and the environment; the closed economy, meanwhile, sets out to encompass social and environmental aspects, but to convince it must paint a world of common agendas, and so deny contestation and differences in social and economic capital. While the first hides by appealing to individual desires at whatever cost, the second hides by celebrating group consensus. How people negotiate between these things is the stuff of the politics of everyday life.

The Politics of Local Food Consumption

We have moved from a discussion of the closed system of the body to personal relations, real or imagined, that operate at varying degrees of intimacy and distance. The protagonists in our case studies share a concern that economic relations between consumers and producers should also be social relations. This is thought to apply to food more than any other form of commodity exchange. In this, people imagine themselves to buck the trend in the mainstream open economy towards impersonality. Whereas some forms of ethical consumption take place at a distance, as is the case with fair trade in tropical agricultural products, there has been increasing emphasis on 'the local' – a defined, hence closed, social space in which 'better' exchanges are thought to take place.

Two key issues emerge here. The first concerns what 'scale' of localisation is imagined; consumers draw their boundaries in different ways at different times. The retail outlet run by the consumer cooperative La Ortiga in Seville gives precedence to local products, but many of the processed organic foods come from the north of the country, particularly Catalonia, and products from further afield are stocked if they are fair trade ones. In the UK case study of farmers' markets, the stipulation is that producers should come from within a 30 to 40 mile radius; in a US study, this distance extends to 150 miles, which makes no sense in terms of understandings of the local in a European context (Okura Gagné 2011: 284). In areas around the small towns featured in our case studies it is easy to imagine a closed economy, and one of the towns, Lewes, has its own currency to substantiate that. But people also draw wider political and cultural boundaries: they prefer British meat or celebrate French cheese.

This brings us to the second issue: the range of ideas about what the local entails. In some discussions, such as those initiated by the Transition Town movement, the turn to localism is not just preferred, it is imagined as increasingly necessary. As oil output peaks, preparations are underway for a 'post-oil economy' based on a model of local self-sufficiency. Under the new rules imagined under oil scarcity, exotic foods will be unavailable and will be replaced by local substitutes or, as one participant in Transition Town Lewes explained, non-local foods may have to be brought in by pack-horse in the post-fossil-fuel world.

This more political kind of localism differs from that which looks to foster personal social relations between producers and consumers, and differs again from the cultural appeal of specialist products.

While there are a range of reasons consumers prefer local exchanges, they are all responses at some level to the opening up of the economy under neoliberal, capitalist globalisation. This is what they share and it is what allows the disparate agendas and practices to look related. That is, they all represent commentaries on the open economy. Analysing this means identifying different advantages of localism, which aspect of the open economy people are responding to, and how this frames consumer practices.

The first perceived advantage of local exchange is environmental, particularly in relation to food miles. Supermarkets have become adept at supplying foods that cannot be produced locally or have a short season – with indicators showing a particularly sharp rise in air freight since 1992 (DEFRA 2011). Food miles add variety to consumer choice, and are also attractive on price. Foods are transported long distances because they can be sourced more cheaply on the other side of the world. Cheap products appeal to consumer thrift. People who shop for local produce try to reduce the use of non-renewable energy and counteract the profligate use of energy in the mainstream food industry. The built environment is another concern; as high streets are now dominated by chain stores, they all look increasingly similar, or worse, supermarkets turn towns inside out and centres become 'ghost towns' Much of the local food movement in Britain is framed in opposition to big, multinational supermarket chains. Using local small businesses then becomes about protecting the authentic character of towns. Lastly, some shoppers associate local food with small-scale, mixed farming, held to be more sustainable than extensive agriculture driven by downward pressures on prices. These farmers are identified as endangered by the price squeeze on agricultural produce and as worthy of support.

This introduces a second thing the people in our case studies want when they consume food – they want the relation between what is produced and what is consumed to be a social relation. They want to invert the open economy or the mainstream market which champions formal impersonal relations; they are anti-capitalist in intent. Often there is a kind of nostalgia for a lost rurality. In a rapidly changing urban environment, people look to traditions of rural village life based around face-to-face relations. This is as true in Spain as it is in the UK, though in the latter case urbanisation took place much earlier, and links to a lost rural world exist more as images and ideas to be consumed than lived social relations (Burchardt 2002; Williams 1973).

Farmers' markets are an obvious example where consumers seek interaction with certain kinds of producers. The sellers at these venues are attractive figures because they are understood to produce at a small scale, to run family businesses formed around closed relations; they produce the products they sell themselves, and they have the special skills and resource bases to create foods with special qualities. Of course those special qualities can also command a premium in terms of market price, which makes them amenable to profit-seeking and entrepreneurial activity, and leads to class-based inflections to their customer profile. This is certainly an element in all the cases documented in this book, but there is also a range of ideas and trajectories. So while there is a common project to construct local economies outside the open, capitalist system, how this is imagined is an outcome of particular economic and political backgrounds. In Spain there is a radical political edge to the tradition of the *pueblo*. Sharpened by socialist and anarchist traditions and experiences of extreme poverty and famine, the *pueblo* needed to be able to feed itself and did so through dense networks of kin, neighbours and friends producing, exchanging and consuming food from kitchen gardens, as well as through locally supplied retail distribution. In the UK there are similar anti-capitalist elements, in allotments, for example, and in the remnants, though recently revived, of localised distribution networks. Nevertheless, in the UK the big retailers control a larger slice of food distribution, and alternative food is a much more middle-class concern focusing on specialist products.

Here we come to the issues around quality. The third thing consumers are looking for is products that are special because of their origins. People are concerned with different aspects of a product's origins: how it was produced and who produced it, what went in to it and how this varies depending on the production regime and, particularly for food, the ecological conditions under which it was produced. In many respects this is the nub of the matter because knowledge of the origin of things and how this is reflected in their qualities presupposes a closed economy, however that is idealised.

This is the nub of the matter in another sense. In recent years there has been a dramatic rise in products that are identified and valued as traditional, local, authentic or artisanal. These typical products are held to have special characteristics due to the combination of local raw materials and traditional production techniques (Treager 2003: 91). Such special qualities are then held to have specific benefits for marginal rural areas: higher farm-gate prices, skilled employment and more sustainable environments, as well as supplying tastier, healthier food to consumers. With its origins in the Mediterranean region, and particularly French notions of *cru* and *terroir*, typically local products are increasingly visible (Pratt 2007; Treager 2003). The issue here is the process of regulation and certification in relation to what in Europe are termed Products with Designated Origins (PDOs) and Protected Geographical Indication (PGI). Three of the countries in our case studies top the tables for such products. In 2008, Italy had 165 listed, France had 156 and Spain had 110; the UK, meanwhile, had 29. This is clearly not the full picture though, because an inventory carried out in the 1990s recorded almost 400 traditional foods in the UK, and about 3,000 traditional producers (Treager 2003: 93). What is fascinating about these products is not so much their popularity or visibility in Sunday supplements, but the circulation through the open economy that certification allows. The products are deemed to have authentic qualities that demand recognition, but the European Union regulations that stipulate the terms and conditions of compliance are precisely what allows the values we have identified with a closed economy to circulate in the open system.

Conclusion

This chapter has identified two contrasting ways consumers think about and act in the food economy, which we have termed open and closed to signal their principle characteristics. Tensions between the two are inevitable because they overlap and coexist in many if not all contexts. This is then experienced as a social and political problem that people have to negotiate in their everyday lives. This problem is intractable because what people are reacting to is the very nature of the open economy's ability to rupture the relation between people as producers, or makers and growers, and people as consumers or users of products. There is enormous variety in the way different actors deal with the tensions between open and closed and its attendant problems. Another way to say this is that there are different ways of dealing with rupture.

At one extreme, the mainstream retail response is to exercise ever greater governance by introducing systems of traceability through the supply chain. Either companies absorb the costs of administering these systems as the price of maintaining consumer confidence, or they pay for it by passing on the costs to customers in the form of premiums. For the latter to succeed, they need to rely on consumers to pay extra for information about the farmer, the production technique or the heritage of the product. Whether they absorb the cost or pass it on will partly depend on what their competitors do, and this then becomes an index of what the market will bear as an acceptable price. In this way, it is a response that works within the open economy. The systems of traceability and governance, as well as attendant premiums, rely on certification systems which allow the products to circulate with value-added in the open economy.

Middle-class consumers provide the target market for these kinds of certified products, and we know this is a relatively successful ploy. In 1999, a MORI poll recorded only one third of people as organic shoppers, predominantly middle-aged women from higher social classes (Cottingham & Winkler 2007: 30). One possible reason such people buy organics and frequent farmers' markets, and an undoubted outcome, is to reaffirm class distinctions (Bourdieu 1984).

There is further evidence that shows a broadening of the appeal of these products across social classes. Cottingham and Winkler (2007: 30) tell us that by 2006 two-thirds of people were buying organic food, an increase of some 30 per cent from the 1999 figure, with the increase coming from low-income customers. The fact that these figures have declined since the onset of the economic recession does not detract from the general point that a widening clientele is an outcome of processes of emulation. If this is accepted, then consumption of organic food is reduced to objective class categories. Actually, at one level this is correct: organic food is a premium product for discerning middle-class customers.

Fortunately, from our perspective, that is not the end of the story. We can see this by looking at the subjective values, motivations and goals of consumers and retailers, which lie alongside and trump class issues, at least in the minds of protagonists. For example, as we shall see, in some situations organic certification is understood as peripheral or even a distraction from the real issues behind alternative food provision. In these cases, class seems much less visible and relevant than in the Anglo-Saxon world. This is so, for example, in Andalusia, where organics are often subsumed under local and regional food, which has long meant resilience for the *pueblo*. This local food economy and food security was built around and required dense social relations with kin and neighbours. Today it also means sustaining political networks of exchange between participants across provinces and regions.

For many, the initiatives are therefore about politics and human relationships rather than class or certification. One organic shopkeeper in Spain cited a single mother who worked as a road sweeper as one of his most committed customers. For him it is a question of political conviction, what one chooses to spend money on, not economic capital. It is certainly true that people in our case studies, some of whom are middle class, are looking for or have been forced by economic crisis into a more political response. The key point is that there are too many variations across the cases in the levels of commitment and in the class profiles of consumers to generalise. Nevertheless, one common theme is conscious rejection of the open economy and support for alternative systems of production and consumption.

These alternatives are creative initiatives, in that they try to generate direct connections between the maker/producer and the user/consumer. The initiatives often seek to bypass class by reaching out to low-income groups as a deliberate policy, and in achieving that goal certification systems can be an expense and therefore a hindrance. Indeed, strategies and policies are largely evaluated in terms of the extent they succeed in connecting people as producers with people as consumers. If they fail to sustain that connection, they tend to move sideways by starting new initiatives with different criteria: local food is the new organic.

The aim, then, is to explore the values that people hold in relation to food and the social forms that they seek to reproduce through systems of provision. These values vary according to context, and while many consumers are happy with mainstream provision, the consumption practices we are concerned with draw on the alternative aspects of food we have associated with closed systems. For some producers, alternative consumption is a window of opportunity. They can draw on the values of closure to generate more money value, and compete in the open economy. For others it is much more of a political project of escape from the terms imposed by the open market.

The effort to separate out the two interwoven strands of open and closed is only fruitful when we look at the complex ways they feed into one another. The case studies which follow therefore document how different people in different ways negotiate the social values they identify with a closed system, and the requirements and possibilities of the open economy.

4

Tuscany, Italy

Jeff Pratt

The second chapter outlined in general terms how small farmers lost their autonomy, and often their whole livelihoods, through revolutions in farming, retailing and global markets. This chapter sets out the practical impact of these changes for farmers in southern Tuscany. It draws on long-term research to look at the way these processes developed over time, and the various strategies which they developed to resist them. Each of our case studies will illuminate a different set of responses to revolutions in the food chain and a different way in which the relationship between open and closed economic strategies unfold. The road taken depends on agrarian history, political traditions, food cultures and other local factors which make specific kinds of economic activity possible. Amongst the households described here, one group developed a socialist organisation, collectivising their land and machinery in order to achieve economies of scale and continue to generate a livelihood in a radically changed world. This was a partial success, but 20 years later, with growing market pressures, new strategies had to be developed. They found these in a combination of reviving older farming practice and the direct sale of specialist, quality products, a market which had opened up because of wider changes to the Tuscan economy.

This chapter follows a historical sequence. Until 50 years ago, these families were share-croppers and lived in a kind of dual economy. Half of what they produced was taken by the landlord and sold to feed townspeople, but the farmers themselves lived in a virtually moneyless world, and the farm was organised around their subsistence needs. After the land reform of the 1950s they began to produce much more for the market, and general competitive pressures began to operate. Incomes were squeezed and the existing economic relationships disintegrated. Farms became more detached from the local economy, they provided a declining portion of local food, and the integration of previous farming practice was fragmented into separate sectors of production.

Twenty years ago, as the squeeze on incomes tightened, many of the younger generation joined the rural exodus while others found employment on the big wine estates. A minority tried other strategies, first through the development of cooperatives to reduce costs and pool labour, secondly through the production of organics and local specialist foods, which were sold direct through farmers' markets and rural tourism. Tuscany has invested heavily in tourism, in its specialist wines and olive oil, and in its heritage industry. This obviously has opened possibilities which cannot be replicated everywhere else, but it does not guarantee success. The British humorist Saki once wrote, 'The people of Crete unfortunately make more history than they can consume locally', and in Tuscany there is currently a similar over-supply of history, landscapes and expensive wines. Consumption is the key word, and the last part of the story explores some of the paradoxes which occur when these qualities become commodities.

The *Mezzadria*

Southern Tuscany is dotted with small towns, usually built on hill tops and numbering only a few thousand inhabitants. They are surrounded by an area of cultivation known as the *contado*, farms with olive groves, vineyards and plough land. Further out, and never cleared for cultivation, are long stretches of woodland. These celebrated landscapes of Tuscany are mostly the creation of share-croppers, who since the medieval period terraced and planted the hills and valleys as part of a land-tenure system known as the *mezzadria*. The farming households were joint families of three generations in depth, working and eating together under the authority of the patriarchal head, the *capoccia*. Brothers stayed together and were joined by their wives until the resources of the farm were insufficient for the labour available, and one of the brothers moved out. Ten people on a farm was normal, and some families numbered over thirty. The landlord could evict a family in the autumn, or move them to another farm, so mobility was common even if stability of residence was generally preferred.

Even in 1970 there was one family in the Val d'Orcia who had been share-croppers in the same place for more than 200 years, all in a classic stone farmhouse, with livestock on the ground floor and above that a big kitchen with the bedrooms opening off it.

Landlords were all urban based, whether aristocratic families with major estates or small-town merchants and lawyers who owned just a few farms. They received half of everything produced, to feed their own families and provide a surplus for sale. In that way the *contado* fed the town: there were specialist crops and some external trade, but generally the town provisioned itself from its dependent farmland. It also contained a full range of artisanal activities: in that way, the *comune* (administrative district) has many of the features of what we have called a closed system. There was also a well-developed pride in local autonomy, going back to medieval times, and the notion that in a 'true centre' a civilised person could meet all their material and cultural needs locally. This, of course, was a view of the world articulated by an urban elite.

By contrast, the lives of share-croppers evolved in a purely rural setting, with a complex division of labour built around food production. Accounts from any period up until the 1960s reveal the astonishing variety of land use. They kept oxen for ploughing and haulage, cattle, sheep and pigs, as well as rabbits and domestic fowl. They grew cereals and fodder crops, vines and olives, fruit and vegetables. In places there were also specialist crops: flax, cotton, tobacco, saffron, and mulberries for silk worms. This was a highly intensive system, optimising variations in soils and slopes, as well as using a form of intercropping known as *la cultura promiscua*. Overall, families had not just their own food, but almost everything else as well: firewood, building materials and much of the raw material for clothing.

Contact with markets was extremely limited; in fact, normally the landlord only allowed the *capoccia* permission to go to town and sell something. Small surpluses provided money for salt, boots and a few extras, larger expenditure (mostly furniture and other goods at marriage) depended on the landlord making up the annual accounts and releasing cash. In one interview, a former share-cropper remembered how his landlord remarked that 'good *mezzadri* need good health, much work and no money'. This encapsulates the dominant ethic, that these farmers should be isolated from the market and as close as possible to autarky – provisioning themselves from the land and through their own labour. Such views also had a political dimension, that the share-croppers of central Italy were thought to be docile, at least compared to rural labourers elsewhere, and that this was best achieved by keeping them isolated from towns and marketplaces. The most unexpected changes could become a threat to the social order, as shown by this comment from Marchese Incontri in 1925: 'the patriarchal calm of the Tuscan countryside has been spoilt by the bicycle and the circulation of newspapers, which have irremediably broken the isolation of the peasants, thus destroying their characteristic ingenuousness, docility, diligence and parsimony' (quoted in Pazzagli 1979: 109).

The Second World War shook up the calm for good. The *mezzadria* system had been a bulwark of Mussolini's fascist regime, while many share-croppers had been involved in the resistance movement. Throughout the 1940s and 1950s, they agitated for a reform programme which would give 'the land to those who work it'. The struggle was led by the Italian Communist Party, and though reform when it came was highly limited, it was a striking period of social emancipation, one which revealed the gap between rural and urban living. The *mezzadria* was bankrupt, with its farmers living without piped water, electricity, access to education or contemporary medicine, and on farms devoid of investment. Where there was no land reform, as in the Chianti, many of the large estates switched to specialist wine production with wage labour, freeing up farmhouses for north European settlers. A few share-croppers were able to buy out their farms, but for the rest it was a period of emigration, with many provinces losing half their rural population in one generation.

Land Reform

The land-reform programme did cover southern Tuscany, and I shall concentrate on this area, particularly a small district on the flanks of Monte Amiata in the south of the province of Siena, and neighbouring communes in the province of Grosseto. Here, in the 1950s, landlords were expropriated with indemnity and share-croppers were granted property rights to about half of their land, depending on the number of adult men in the family. New farms, with their white, box-like houses, were created for newcomers, generally rural labourers from the mountains. What kind of farming system did the land-reform agency try to create? It seems to have been an unstable mix of two visions. One was derived from Catholic social doctrine, in tune with the ruling Christian Democrat government, eulogising the patriarchal farming household and its relationship to the land and to God. There was a strong emphasis on self-sufficiency: each farm got a bread oven and its small cattle stall. They also dug irrigation ponds and planted up new areas of olives, vines and fruit where they were missing on the new farms. These crops might generate surpluses for sale, but the first priority was that they should feed the family, through manual labour.

Alongside this model was the ambition to bring these farmers into a market economy. A cash income for these farmers was generated out of arable farming, and this was mechanised, using centralised cooperatives with a pool of machinery. Land was deep ploughed for the first time, then seeded, fertilised and harvested in operations planned and executed by these service cooperatives. All the farmer had to do was tell the cooperative to set this in motion; then, at the end of the year, they received the accounts and banked the balance between what the cooperative charged and what it received for the cereals it had sold.

Living standards rose rapidly as a result of these investments, as did state subsidies available for agriculture. Rural schools were built; by the 1960s most farms had piped water and by the 1970s electricity. This made possible more urban-style kitchens and bathrooms, and was one of the factors which made these households 'consumers' in the modern sense: they went shopping. This in turn obviously meant greater dependence on a cash income rather than subsistence, even if food still represented about half of an average household budget in this period. The dual structure began to evolve almost as soon as it was created. By the 1960s, farmers began to buy their own tractors, in part because they wanted them for general farm operations, in part because they wanted more independence. The problem was that the farms were really too small (15 to 20 hectares) to justify this kind of individual mechanisation. Above all, this shift in the household economy made them much more specialised and vulnerable to changes in global commodity markets. The effects vary from sector to sector, and there have been good times and bad times, but overall most farmers are struggling to survive, and pessimistic about the future for their children.

We can see what this means in practice by looking at different kinds of market and how they evolve, starting with subsistence produce from the barnyard and the kitchen garden. Since the share-cropping days, this produce had always fed the family and provided small surpluses which women sold. Each farm had regular customers who passed by from nearby towns and villages, or who showed up maybe only once a year for a few demijohns of wine and a supply of olive oil. This has all been hit hard by the opening of supermarkets, where people now buy most of their food, and by government regulation. Here is Graziella in the yard of her farm at Porrona in Grosseto province, venting her frustration in 1987:

> The farm is too small. It was always too small, back at the time of the reform. They gave us farms that were too small. It has now got worse, you cannot sell anything. We have given up on cattle. But it is the same with pigs and eggs. At least here at home I could do something. We always keep chickens and a pig or two for the family, then we could sell a few, to do the shopping. People used to come round these farms and buy pigs and eggs but they do not come any more. And now you need a certificate to sell eggs. Now we just throw the eggs to the dogs. It is a disaster.

As Graziella remarks, they shut their cattle byre just like virtually every other farm in the neighbourhood. This is another market which has been transformed. Once it was entirely local, with a handful of buyers touring the district taking the steers for slaughter. Then the meat supply chain was reorganised, consolidated around wholesale suppliers in places like Bologna, 150 miles away, where they could find meat at lower prices. Local farmers had not previously had to compete with the most specialist livestock producers in Europe. One problem was that the Tuscan breed, the magnificent white Chianina cattle, did not put on weight very quickly. One group of farmers at Montenero tried to counter this by opening a collective cattle stall which they stocked, not with Chianina, but with fast-fattening breeds bought in from France and Poland. The next problem was feed. They had hay but very little irrigated land to grow specialist fodder crops, so they bought in high-protein animal feed. They went bankrupt in five years, though by that point they were largely detached from a local farming economy: they were making money only for French cattle breeders, American soya growers and the banks.

A similar process has been unfolding in other sectors. Customers for table wine declined in the 1980s and many vineyards were grubbed up. Then in the late 1990s many farmers in Grosseto province were tempted to go up market, fascinated by the spectacular prices for wines like Brunello across the Orcia River at Montalcino. They obtained new DOC (*denominazione di origine controllata*) classifications and began bottling. Ten years later, a few have made a success of it, but the majority have failed, the result of indiscriminate legislation and a glut at the top end of the market. In 2012, even famous Tuscan producers had cellars full of very expensive unsold wine. A similar trajectory holds for olive oil: families used to buy a year's supply of oil for cooking and bottling vegetables; now they tend to go to supermarkets. The situation is not quite so dire as for wine, since there is a price premium for good Tuscan oil, something which has also been recognised in place-of-origin legislation. But this is still a delicate situation given that the old form of direct sales has declined, and that the generic market for olive oil is provided by supermarkets which can source from Spain at half the price.

The final and most important crop is wheat. Village-based cooperatives have made massive investments in the technology of modern cereal farming: from self-levelling combine harvesters to computer-controlled grain silos. They specialised in hard (durum) wheat, and over the last decades there have been some boom years, especially in the 1980s, with high prices and EU subsidies. However, this is one of the most volatile of all agricultural markets, integrated at a global level and subject to spikes and slumps. Even fluctuations in the exchange rate of the Canadian dollar have a direct impact on Tuscan livelihoods. Because growing wheat involves high and easily calculated costs, farmers are keenly aware of their 'break even' point in terms of yields and prices. In the last few years many have been below that point, partly because yields are low, partly because even when the price of wheat has tumbled the price of diesel oil and fertilisers have not. This is what is really worrying the farmers and cooperative managers. A generation ago they shrugged and ploughed the fields even in the bad years, 'so as not to let the land go to waste'. Now they wonder whether the price squeeze will ever relax its grip, and if there is any future for this kind of arable farming in the district.

In many cases these farms now consist of a cereals monoculture, with a kitchen garden and a few chickens attached. Since the 1980s, this simplification has for many been driven by the attraction of wage labour on the nearby wine estates of Montalcino. In the last decade, even this source of income has diminished. The estates still hire specialist tractor drivers, but for the pruning and harvest they now employ squads of Bangladeshi, Kurdish and Moroccan workers. The internal working of these employment contracts is hard to penetrate, but locals refer to them as *caporalato*, a term used for the ruthless gang-labour system of the last century. At any rate, it is not just goods which travel and transform this rural society – there is an international labour market too.

Fifty years ago in this district was to be found an integrated pattern of farming, based largely on manual labour, creating a local food system. All this has now been blown open. Farmers no longer practise mixed cultivation, the farms are now highly specialised and depend on expensive industrial inputs. Even when they are not monocultures, the different sectors of activity have scarcely any connection with each other or with the rest of the local economy. Each sector is integrated into widening international markets, and has to be able to compete with prices set by those with the best natural conditions, most efficient technology or cheapest labour. On these hills, that challenge has become increasingly difficult for both cereals and livestock. It is a difficult period, and a researcher encounters resentment, puzzlement and silences. Farmers regret the destruction of the investments made through the hard labour of their parents. They worry about the long-term future of their farms, whose soils are no longer renewed through livestock and crop rotation. They question the choice to seek off-farm employment, which should never be at the expense of all farming activity: 'One brother should always stay and cultivate the land because it was left by the parents. But the best yields come from wage work'. They make troubled comparisons with the past: 'When I was a kid, this farm fed a family of 14. Now there are four of us, and I still have to go and find some wage-work (*a opera*) to make ends meet'.

It is true that living standards have risen substantially over this period, even if farmers habitually talk much more about disasters than the good years. However, even this generalisation needs serious qualification. Incomes have grown for a variety of reasons, including the growth of a welfare state, overt and covert farm subsidies of myriad kinds, and recourse to off-farm employment. Above all, we need to remember that the productive resources of this land support fewer and fewer people: current farmers are survivors from a continuous rural exodus which has seen their brothers and sisters leave for higher incomes in the city.

Working out how these farmers achieve what they consider an acceptable livelihood and when they give up is an intricate matter. For capitalist enterprises, like the wine estates, this is relatively simple: there is a bottom line and it is calculated by subtracting production costs from sales figures. The production costs include those generated by the purchase of land, depreciation (of machinery, buildings and other fixed assets), materials costs, and labour. For a farming household, the situation looks rather different. Land is inherited. Machinery purchases and other investments are made in good years when the cash flows in, but after that they are not thought of as an annual cost. If the tractor is paid for, you might as well use it. Labour is not a cost, it is what you do in life. What you get back materially from this work has always been unpredictable – that is the nature of traditional farming. A bad run which puts a capitalist firm out of business can be weathered for some time, the family makes no investments, tightens its belt and accepts a lower standard of living: 'We break even because we do not count our sweat'. There are two things which precipitate a major change of direction. One is when the unavoidable material costs of production exceed the returns: that has happened recently for cereals, with rocketing prices for fertilisers and fuel; it also happened to the ill-fated beef-fattening cooperative. The other is when off-farm employment provides a direct point of comparison with farm incomes, and equations of time, work and money permeate everyday lives. This generates serious reflection about what the family gets out of farming, materially and intangibly, and what is at stake if they sell up.

There is no doubt these are tough times. In 2011 I talked to many old acquaintances – cooperative managers who are well placed to see the aggregate trends, as well as to individual farmers – and they were all profoundly uncertain about the future. There are, however, glimmers of light from alternative patterns. Some were visible 20 years ago: families who had maintained mixed farming practices, with a wealth of food for home consumption and some local customer base for their surpluses. They did not think about their farms in terms of different sectors, each of which had to compete in specialist markets, but in terms of maintaining the productivity of their land in an integral way. By holding on to livestock, for example, they had kept up the older pattern of crop rotation, which also cut their costs when they did choose to plant cereals. When asked to comment on this more integrated practice, they said, '*tutto fa*', 'everything contributes'. Their neighbours thought they were worthy of respect but distinctly old fashioned. They certainly did not have the highest incomes in the district at the time, but they did seem more autonomous. In the second part of this chapter I shall concentrate on one group of farmers who maintained features of traditional husbandry while experimenting with just about everything.

Ripa d'Orcia

Farmers like to be their own masters and are possessive about their land: jealous, as they say in Italian. So it was a surprise to come across a group of families who had pooled their land and their machinery, and were working it collectively. There were unusual circumstances. The households had all been *mezzadri* on a particularly run-down estate situated on a long ridge of land to the north of Monte Amiata in Siena province. The struggle for reform was a long and bitter collective struggle: they even besieged the castle from which the estate was run, skirmishing with police for twelve days. Finally, in 1955, 15 families gained ownership of their farms.

Collective action continued after the reform, and in 1970 they jointly purchased a combine harvester. Next up came the tractors: was it worth re-equipping individually? Or should they buy machinery collectively, in which case they would have to work the land as one unit. There followed nearly a decade of difficult discussions. The regional research agency surveyed their farms and their labour, the Communist Party and its farmers' union tried to help, but they had no models to go by. Eventually, in 1980, enough loans came through for them to start, and nine of the fifteen households who had taken part in the discussions did pool their resources in an experiment which lasted more than 20 years.

There is no doubt of the radical political edge to this cooperative, or its historical roots. After all, if your landlord has been expropriated following a campaign based on the slogan 'the land belongs to those who work it', you have asked serious questions about dominant views on property. It is more unusual to continue those questions after you have won. Delfo, one of the founding members, told me the land reform was all wrong: '[I]t should never have given property rights; that just starts up capitalism again, with people selling up and big estates emerging. It should only have been rental agreements, conditional on you staying and working'. Essentially, the cooperative was created in order to improve the viability of small farms and to create a future for their children, at a time when everything around them was going in the opposite direction. One statistic shows the scale of the changes: in 1951, 64 per cent of the population of Siena province worked in agriculture; by 1981 it was 12 per cent, of whom half were wage labourers on the wine estates.

The cooperative made an immediate impact on all its members' lives. It is a revealing experiment, in its successes, its failures and its inadvertent consequences. Each family kept 3 hectares of private land for subsistence activities, vegetables, pigs, chickens and vines. They pledged the rest of their land to the cooperative, for 20 years in the first instance, making about 240 hectares in total. They built and stocked a cattle byre, established a flock of 700 sheep with a milking parlour, and bought more powerful tractors. They aimed to achieve economies of scale in some activities, use the land better by matching terrain and soil to use, and above all to develop their own skills more productively. One person looked after the cattle, another did most of the mechanical repairs, others did the milking. Elvio became a full-time shepherd and wants nothing else. He told me about the complicated life of his father, how in the height of summer he used to get up at 2 AM to feed the oxen for a long working day, and how he went to bed still worrying about all the tasks yet to be done. Elvio does not have to worry about feeding the cattle, or dealing with a broken-down tractor. He walks the sheep, from dawn to dusk, patiently finding and husbanding their feed, from the pasture to the woods to the wheat spilt at harvest in the open fields.

The lives of farming women changed even more dramatically than those of the men. They were used to working in isolation, generally on activities which were not generating cash incomes. After the reorganisation they tended to work together, milking the sheep morning and evening, or picking olives. Above all, this work is paid at the same level as everything else, it being a founding principle of the cooperative that all work had the same remuneration. In many households women were working longer hours, and so earning more than the men, and given the organisation of the cooperative they had a clear voice in the decisions which affected their livelihoods. There was a notable break from the old patriarchal patterns found in the *mezzadria*, and on many of the land-reform farms.

One of the most complicated tasks was agreeing how to calculate their incomes. All the other cooperatives around paid out a fixed wage based on union rates, but at La Ripa incomes would depend directly on how much money the cooperative made. At the end of the year, all the operating costs are subtracted from the sales receipts, and the balance is divided: 20 per cent according to the value of the land each member has pledged, 80 per cent according to the labour provided. An hourly rate is calculated according to the total number of hours worked by all members. Research done before the cooperative started had come up with the depressing fact that if these family farmers costed their time, they were working for a quarter of the hourly rate for agricultural labourers. It set up a comparison which was in some ways misleading, but it did provide a yardstick for measuring progress, and by the end of the 1980s they had doubled their 'wage rates'.

The cooperative continued to evolve in the 1990s: they planted olives and bought more land, principally from three of the families which had stayed outside at the cooperative's formation but failed to survive. They turned the new farmhouses they had acquired into rentable accommodation for an experiment in rural tourism. There were failures, precipitated by the same economic forces we encountered earlier. After more than a decade, the Chianina cattle were sold off but they kept the sheep. They reduced the area planted with cereals to the most fertile land, and on a longer rotation, two years out of five.

Some of the next generation joined the cooperative, but significantly the children in some families did not. It was partly because the comparison with hourly rates in other sectors was out there and visible, partly because farm work of this kind is flexible and unpredictable. The proposal to create an eight-hour day for the young provoked a major disagreement and was rejected. Partly this was on practical grounds: do you go home at 5 o'clock if a cow is giving birth, or the weather finally allows you to be harvesting? The phrase 'an eight-hour day' I think also struck a nerve, that what they were doing would 'collapse' into wage labour. They had created a complex connection between time, work and money, but as farmers the question of whether work, as measured by the clock, actually generated money was very different from a wage-labour contract. It depended first on their dedication and skill, and secondly on the unpredictable play of weather and markets.

To give an idea of this raw nerve, here is Loreno Mascelloni talking about a neighbour who sold up to work as a wage labourer in the building trade:

> I could not do that. Man's initiative finishes that way, you become an object, it is no longer you who decide, who creates projects, who makes plans, who develops your own personality or your will. For a lifetime you have to do what you are told. You sell your labour to others. I do not feel like working like that, as if the principal thing in life is getting money, taking the attitude that I don't care whether the job is done well or badly, they pay you all the same.

Farming, as Loreno and other cooperative members understand it, is incompatible with fixed hours, and with much of the leisure activity that is associated with a consumer society. Elvio, who is 'very attached to the four walls of this house', gets fired up about the problem of taking days off: 'It is not just the problem of penning up the sheep. Here we have rabbits and chickens and a kitchen garden. What happens to them if we go away for three days? We country people are used to keeping these things. It is useless thinking that we can lead a life like other people'.

The Present

After 20 years, the cooperative started to accumulate problems. Some of them were technical and bureaucratic, since the members were both independent farmers and employees, and this created difficulties on everything from pensions to subsidies. Others were internal: some of the children of the founding members joined the cooperative and some stayed out, so as time went by the original equality between families became unbalanced. Renewing the membership and social capital of production cooperatives is always complicated, especially so if it involves families. By 2010, most of the original organisation was effectively at an end. They continued to own substantial areas of land and two farmhouses collectively, but mostly they worked it individually, renting specific portions from the cooperative according to their labour resources and interests. Does that make the cooperative a failure? Not entirely, for two reasons. First, they are still there and farming, whereas most of their neighbours on that hillside who did not join have sold up. Secondly, their current farming practices have their roots in the cooperative experience, and they carry forward the reorganisation of land use and the specialist skills developed then.

The cooperative members still operate a mixed farming system, and the key to this are the sheep. The milk is sold to a dairy making pecorino cheese in the Val d'Orcia, one which won a prize a few years back for the best pecorino in Italy. They graze on pasture sown with alfalfa and clover, which provides nitrogen for the soil, and in addition the sheep provide manure. This means that they can plant cereals on the best land, and after a long rotation. For a variety of reasons, including costs, they are moving to a closed system in maintaining soil fertility. Farmers without livestock make heavy use of nitrogen fertilisers, and have a problem finding any profitable rotation crop.

However, the most interesting story is not the survival of extensive farming, but what happened to the land set aside for home consumption. This had always been outside the cooperative accounting system and was worked by all the family in spare moments, especially the elderly, who continued to contribute to the domestic economy. As we saw from Elvio, it was central to their understanding of rural life. The development of tourism gave it a new lease of life. *Agriturismo* is a rural bed-and-breakfast scheme particularly favoured in Tuscany as a way of boosting rural incomes, and of decentralising tourist flows in the region. Loans are available for converting farm buildings, though there are regulations which stipulate that tourist income should not exceed that from farming. The cooperative had pioneered this locally, and it then took off individually as many of the families opened up accommodation in their houses.

The result was a boost for the more artisanal and labour-intensive forms of farming. Tourists consume fruit, vegetables and eggs while they are staying, and buy other goods to take home. Families began to bottle wine and olive oil; some also sold honey or pecorino cheese they obtained back from the dairy. These direct sales to visitors from Italy and northern Europe were further boosted by a growing reputation for quality, and there are increased sales to families and restaurants in the locality. One farmer has a passion for wine and made a long-term investment in new vineyards, a cellar and DOC recognition. Four farms have been certified as organic, and two more are making the transition. There are an increasing number of farmers' markets held in provincial small towns, and one family sold organic chickpeas, lentils, beans and spelt every weekend, until ill health forced them to retire. Others are exploring this option, while even without certification these farmers can sell all their olive oil directly to consumers.

Overall, the farms at La Ripa have reduced or abandoned producing those products where international competition is at its fiercest – such as wheat, beef and lamb – and now concentrate on specialised, 'quality' foods, some certified (DOC, organic) and some not. This in turn has come about through the creation and re-creation of different markets, nearly all of them dependent on direct links to consumers. At one level the changes are very simple: they have got out of those sectors where they made a loss and moved into ones where they can still make a living. It appears that these are rational responses to market conditions, and perhaps that is all we need to know to understand what is going on. However I have already suggested that in many circumstances the accountant's bottom line' is not a make or break calculation for these households, while in terms of simple market logic the rational people are those who sold up not those who are still farming. There is more to be said about money, about the other reasons people grow food, and about the relationship between the two, but first we need to say something about consumers.

Food Values

Historically, food is more important in the social and cultural life of Italians than it is, for example, in Britain. Families still generally eat a cooked meal together every day, and when they do there is often a good deal of commentary on the quality and provenance of the ingredients, as well as on the merits of the cooking. At least a million non-farming families grow some of their own food. Italians in general are less remote from rural life than those in the UK or the USA: half of them were still living in the countryside after the Second World War, and many urban families have members who grew up there. Ingredients and recipes are marked by regional and local variation, and each place seems to take pride in its distinctive way of making a vegetable soup or preparing a *ragu*. Of course, all this is evolving. There is a slow decline in cooking from raw materials, some atomisation of family life, a growth in supermarket-based food lines, and the 'culinary unification' of Italy around the likes of pasta and pizza. Nevertheless, people are acutely aware of these changes and discuss them in a way which is familiar from other cultural spheres. There is the 'modern', standardised food world, convenient but anonymous, and there is the traditional world, distinctive, rooted and human. The Slow Food movement, born in Italy, articulates this distinction constantly.

Inevitably, these are generalisations, but they help us understand what is going on when people seek out local foodstuffs and direct contact with producers. Some of the produce from the Ripa farms is sold to people from the neighbouring town who come out to buy olive oil and table wine. They buy it because they like the taste: even for these everyday items there is discernment in people's judgements and a vocabulary to express them. Customers always say it is *genuino*, 'pure', as it should be. The wine is 'just made from grapes' since a good producer can make a wine which is stable without using any chemicals, just by racking it properly at the end of the winter; the olive oil is not blended with inferior oil from elsewhere. They know it is *genuino* because they know where it comes from and who made it. Some townspeople cultivate a network of long-term relationships. Through them they can source food which is both *genuino* and *nostrano*, 'ours': the particular variety of purple artichokes, the small white peaches, and if they are lucky the bottle of homemade *vin santo* ('it has no price'). They value this social landscape as much as they do the knowledge of where to find wild asparagus in the spring and *porcini* mushrooms in the autumn. I think the value is coloured by the fact that this is no longer the normal way of sourcing food: it is now a choice, and a return. Peasants (*contadini*) have gone, the towns no longer obtain the bulk of their food from their rural district (*contado*), and even these customers for local food buy their basics from the supermarket.

Those who stay on the farms for holidays are of course not local. They are from Italian cities, and, since the start of the recession in 2007, increasingly from northern Europe and in search of more luxurious accommodation. At La Ripa we find a mix: professional couples wanting a few quiet days in an iconic landscape, and families having an affordable holiday with the kids. Many come back every year. Nothing is formally organised for them, but if they want a natural history tour or information about farming, it is available. When I went on a tour with them it was a large and inquisitive crowd. They wanted to see the vines which produced the wine they had been drinking, and they asked why they were planted on that particular stony slope, how the micro-climate works on Monte Amiata, and what were the effects of the great frost of 1985. They heard about the problems with the wild boar that infest the woodlands all around, the return of wolves to the district, and a dozen other things about the history of farming in the area, including its social and political history.

Other stories deal with food itself. Once over supper I heard Loreno explaining to his visitors the distinctive features of the pecorino cheese they were eating. It used to be made at home; his mother was considered the most skilled in the district and was always called in to the castle to make cheese for the landlord. Even now that pecorino is made in a dairy, there are other features which make it distinctive. There is the breed of sheep, the timing of the lambing, the pasture. Cheese made in May is best, the fields are full of aromatic plants which give flavour to the milk. The cheese should be kept in a cool room for four to six months, oiled, turned, and ideally wrapped in walnut leaves. At the end of all this, one of those present said, 'Loreno that is wonderful, why don't you write it all down and put it on a label'.

Several things are going on in these encounters. Tourists are connecting to a particular place, in a country which constantly regenerates regional differences and stereotypes. They are participating in the Tuscan experience, a region which over the last 30 years has built a large part of its economy around the heritage industry. Secondly, they are acquiring some knowledge about how food is produced, food they have eaten and will take home, probably to share with friends. They are connecting to the kind of world that was once familiar to most people, but is now hidden by the workings of global food chains. I think this means that their experience tends to connote the past, whether or not they are observing artisan traditions, and this gives it a nostalgic aura. Out of all this the food acquires a history, one which they can retell when returning home with their bottles and jars. They cannot say it is 'our' food, as do local customers, but they do have a story to tell about where and how it was made. This link to production is what makes any commodity 'authentic', and sets it apart from the unknowable origins of mass production and global markets.

There is a third theme which overlaps with the local, the genuine and the authentic. It is 'quality', a rather slippery term in English and Italian. On the one hand, *qualità* simply means the specific characteristics of a thing. If you ask 'what *qualità* is this apple', it means what kind or type of apple is it (a Golden Delicious, for example), just as the word *razza* means what breed of cow or sheep. So when Loreno tells the story about pecorino, some of what he is telling us is what makes it what it is, what distinguishes it from other cheeses. When out of curiosity I took him some English stilton and asked him what he thought, he had very little to say. He could not know whether it was a good or bad stilton, and neither of us knew much about how it was made. However, as in English, *qualità* as distinctiveness shades over into *qualità* as excellence. The argument shifts from distinguishing between two kinds of cheese to a claim that one of them is superior, the basis for a sales pitch that it is worth more. The word quality is everywhere in writings and conversations about food, and its slipperiness alerts us to the fact that something important is going on. Two apparently distinct realms of value overlap here: the romantic values of cultural traditions, distinctiveness and authenticity; and the commercial world of increased profits. The concept of 'quality' makes the shift seamless.

Most people can only obtain these goods by being customers and paying for it – qualities become a quantity of money. Do customers want some proof of quality, provided by a certifying agency external to the transaction, which justifies the price? Within a social network, such as local customers buying direct from a farm, the answer seems to be 'no'. At social gatherings in Tuscany it is very common for someone to arrive with a few bottles of unlabelled wine to be shared and judged. The lack of a label (especially if the wine is from a famous district like

Brunello) is evidence of the quality of a person's social network, as well as an implicit critique of certification and those who rely on it. A local speciality, after all, has to be 'exclusive' to a locality. The play with unlabelled bottles, like the tacit knowledge and the banter of such social occasions, is the other side of the coin, the working out of inclusion. For tourists on the other hand a label may be essential if they need evidence of authenticity that they can take home, shared outside the social network in which it originates.

Conclusion

Producers and their customers need each other, but there are inevitable differences of perspective and interests when they meet. This is found in all market relations, even if the only issue is price. In situations like this in Tuscany, price is not the only issue; instead there is a complex interaction between market processes and the values embedded in the social history of the district, those of household continuity and autonomy, of sustainable farming, as well as various forms of solidarity and inclusion. This is true of both producers and consumers, though it plays out in rather different ways.

Farmers' exposure to market forces is comparatively recent, only beginning after land reform created a commercial cereals sector. The major shift away from subsistence also occurred at this point because these families needed a growing cash income, both to buy the new inputs necessary to farm, and for their own new consumption patterns. Since then they have been forced to give up one sector of production after another in the face of the international integration of agricultural markets. As a result, there has been continued rural out migration, while many of those who stayed simplified their farming operations and rely heavily on off-farm employment. This chapter concentrated on one group (there are others) who took a different road. They were determined to survive as family farmers; in fact the driving force behind the cooperative was to create a viable livelihood for the next generation. This old sense – that each generation is a custodian of the energies of their predecessors – is very strong: they want to leave something. They work long hours, invest heavily and live thriftily: battered cars, work clothes and an occasional week's holiday by the sea. They continue to value a large measure of food self-sufficiency, as something intrinsic to rural life, and the organisation of the cooperative deliberately 'ring fenced' this activity. They developed a socialist organisation, sharing their labour, in order to survive as farmers against the odds. I have tried to indicate why after 20 years it dissolved slowly, and what is its legacy – including the fact that they have survived.

They have concentrated on the direct sale of things which are also part of their own consumption, including specialised foods whose qualities are specific to the area. They may not be able to compete on price with French wheat producers or beef farmers, but they can compete in the supply of milk for *pecorino Val d'Orcia*. Looked at over a 50 year period, it is a curious trajectory. They have not set out to preserve artisan traditions; on the contrary, there has been continuous technological innovation. However, they are selling 'rural experience' to tourists – absolutely unthinkable 50 years ago – and food items which are part of their local diet. They are doing so in a radically changed world, where the significance of the experience and the diet are a counterpoint to modernity and industrialised food. This is also true, to a greater or lesser extent, for their customers and consumers. We find, folded together, the long-term values of a family-farming system, and the financial opportunity of a renewed interest in local markets and specialist products. I am not sure that they continued with these farming systems and their values in order to find those markets; rather, they held on long enough for the world to come and find them.

This plays out in different ways amongst the group of farmers, and some are certainly more opportunistic and entrepreneurial than others. The issue of quality brings out the two faces of this reality. On the one hand, they are continuing a specific production system, whose methods, varieties and flavours are valued, recognised and circulate within local networks. On the other hand, they are part of a larger economic world, sometimes unwillingly, sometimes by design; a world where distinctive qualities become superior foods with high prices, sustained by prizes and publicity. This is the world of the 'Slow Food dilemma', where products which (in reality or in myth) originate outside the base realm of commerce, come to join its circuits, fresh, special, and with much value added. Here local foodstuffs, crafted with love and by definition in short supply, become available to international connoisseurs through tourist guides and the export trade. For this circulation to be possible, there has to be some system of certification, and a label. The paradox is that in the world that the label tries to evoke in its sales pitch, no labels are necessary.

Sources

Much of this chapter is based on anthropological fieldwork in a number of locations in the provinces of Siena and Grosseto in the 1970s and 1980s. In particular, it draws on two projects concerned with rural transformations. In 1984 I conducted research with anthropologists at the University of Siena, who were studying social and cultural themes on northern Monte Amiata. I concentrated on a variety of cooperative movements, and am profoundly grateful to Pietro Clemente for this opportunity, which was also how I first met the farmers at Ripa d'Orcia. In 1987 I received an ESRC grant to study the organisation of the wine estates of Montalcino, and their impact on the neighbouring land-reform farms. This chapter draws on a fuller account of these projects found elsewhere (Pratt 1987, 1994). I have continued to discuss farming problems with friends made during this research every year since.

Further information on the *mezzadria* can be found in Silverman (1975): she researched the system while it was still being practised in Umbria. Italian sources include Clemente (1987) and Giorgetti (1982) on the agrarian history of southern Tuscany. Dickie (2007) is a recent study of Italian food cultures, though older sources such as Camporesi (1993) give a fuller account of rural diets, while Petrini (2001) and Andrews (2008) deal with the Slow Food movement.

5

The Tarn, France

Myriem Naji

In France, the market for organic produce increased by 32 per cent between 2008 and 2010. The distribution circuits are varied, ranging from supermarkets, specialised shops (independent or franchises such as Biocoop) to direct sales on the farm, and through weekly markets and box schemes (Agence BIO 2013). This chapter looks at 30 small-scale organic farmers and artisans of the Tarn district, in the Midi-Pyrénées region of south-west France.

The Midi-Pyrénées is one of the three most important organic farming regions in France, and the chapter explores the motivations, values and strategies of Tarn organic farmers. It focuses particularly on those who are most critical of the commercial logic and 'productivist' practice of mainstream agriculture. They share a strongly articulated desire for autonomy and seek to realise this through developing farming practices that reduce their dependence on the powerful circuits which shape mainstream food chains: agribusiness, supermarkets and state regulations. They embrace ethical, social and anti-capitalist values, but within this group there are different emphases. The chapter is divided into three main parts. The first section situates Tarn organic producers historically and socio-economically. The next section considers the values and practices of these farmers as they struggle for autonomy in production. The third and last part of the chapter focuses on the concerns these farmers have with human relations in labour exchanges, commercialisation activities, knowledge sharing and leisure activities.

The Tarn and Its Farming History

The Midi-Pyrénées region has the second highest rate of population growth in France due to the arrival of French and British pensioners over the past 15 years, and urban dwellers moving to rural areas, the so-called *néo-rurals*. Agricultural and industrial production is the third source of wealth in the region (17 per cent), after tourism (19 per cent) and pensions (28 per cent). Given the low population density, the large number of small farms and the different environments, local people can buy a wide range of foods directly from producers. Intensive production of wheat, sunflowers, soya, maize, rapeseed and pink garlic is found in the valleys of the Agout and the Tarn, whereas mixed farming tends to be practised in the mountainous, hilly or marginal areas: sheep and cattle for milk and meat, pig rearing, aviculture and vegetable growing. There is also a tradition of ewe milk production for the Roquefort companies in the north of the Tarn, while in the Gaillacois there is significant production of quality wine.

Historically, the Tarn has always been a place of refuge and of resistance to the state, with its corollary repression, as experienced by the Cathars in the thirteenth century and Protestants in the seventeenth. There is also a strong tradition of socialism, attached to the important coal, leather and textile industries. Jean Jaurès, a famous socialist leader who defended local miners and small farmers at the beginning of the last century, was originally from the area of Castres in the southern Tarn. At that time, a democratic union of small farmers gave agricultural workers and tenant farmers access to property following the decline of the rural bourgeoisie (Taillefer 1978). This new class of small peasants created a more mixed subsistence farming system oriented towards local markets. Despite being a small landowner class, they were anti-clerical and had socialist sympathies. They were a major force in the local Radical Party and saw to the development of cooperatives and agricultural unions, the *syndicats*.

The Tarn was one of the last departments in France to embrace industrial agriculture and tractors. Indeed, even in the 1960s, rural exodus was slow in the region and small farms survived in great numbers, often thanks to secondary revenues from factory work by worker-peasants (*ouvriers-paysans*) employed in the textile and leather industries. Farmers continued to practise mixed agriculture, feeding their families from the land and selling the surplus in towns. As we shall see, the historic link between towns, the traditional place for markets and employment, and the countryside still exists, albeit in a different form.

Since the 1970s, the Tarn has been a place of re-peasantisation by two categories of farmers attracted by the cheap price of the land left by retiring farmers whose children were not interested in taking it over. One group is made up of young landless children from large farming families in the north of France; the other by young, urban, post-*soixante-huitards* (inheritors of the values of the student protests of May 1968) originating from other regions in France, but also from Holland, Belgium and Germany. These two groups have historically been associated with rather different political cultures, even if with the passing of the generations there have been some crossovers.

The landless children from northern France arrived through a political and religious movement, the Jeunesse Agricole Catholique, and from there fed into the creation of the Confédération Paysanne (known as the Conf). This is now the second largest farmers' union in France, although it is still tiny compared to the powerful FNSEA (Fédération Nationale des Syndicats d'Exploitants Agricoles). The Conf is known for the dismantling of the MacDonald's outlet in the town of Millau (in Aveyron, a neighbouring district), an event organised by its famous leader, José Bové, who has positioned the union as a global peasant movement against neoliberal globalisation (Williams 2008a, 2008b). In its politics it is part of a broader, more radical alliance that includes groups like ATTAC, an organisation that supports the Tobin tax and social and ecological alternatives to neoliberal globalisation. The Conf embraces a set of goals which are shared by many farmers, and increasingly by social movements concerned with the excesses of the agro-food system. It criticises state agricultural policy for letting consumers have cheap food at the expense of farmers, demands that farmers are paid for their work at its just value, and that subsidies are removed or allocated in different ways.

In pursuit of the alternative, the Conf has developed a charter for peasant agriculture (*agriculture paysanne*). This is against 'productivist' industrial agriculture, with its specialisation, standardisation and concentration of resources, and in favour of small-scale farming that respects the environment and is sustainable in the long term (Bové and Dufour 2001: 220). At first sight these goals seem close to organic production, but the situation is more complicated. Many organic farmers have left the Conf because of what they saw as double standards: on the one hand it argues for *agriculture paysanne*, and on the other many of its members are deeply involved in the agro-food system, still practising a conventional or 'integrated' type of production using synthetic inputs, antibiotics and silage. This is one contradiction in the Conf: although it supports organic farming, it cannot afford to lose the support of 'conventional' farmers, who form the majority of the membership of the Conf. Although the Conf is at the forefront of many oppositional claims, with regard to organic farming there are more radical positions.

One of these, which is very active in the Tarn, is Nature et Progrès (N&P), a social movement created in 1974 as the earliest French organic association for farmers and consumers. Today it advocates a return to authentic organic farming. It contests the conventionalisation of their original movement by the state, at both national and European levels, and proposes a participatory certification scheme. N&P attracts the support of many of the second group of farmers who settled in the area, the *néo-rurals*. The heirs of May 1968, these *néo-rurals* reject capitalism, industrialisation and productivism. Like the original *soixante-huitards*, the younger generation of *néo-rurals* who settled in the past five to ten years in the Tarn seek new forms of sociability, egalitarian relations and a relationship to work that is a source of satisfaction and meaning

To further these ambitions, N&P created a yearly ecological and organic fair called Biocybele in the town of Gaillac. In Castres, Albi and Gaillac it also created three weekly organic farmers' markets, called *noctambio* because of their late opening hours. These *noctambio* markets represent a long tradition of producers supplying local food to local consumers, now extended to organic consumption. To compete with the supermarkets, members have a policy of offering as much and as varied a range of foodstuffs as possible in a one-stop shop. In Castres, there are at least three vegetable stalls, three dairy-based stalls (specialising respectively in cow, ewe and goat produce), two bakers and one patisserie, a poultry and egg producer, one pork producer, a veal producer, a wine maker, a flour seller, and a stall offering jam, chutney and dried herbal products. As they aim at diversity, they do not admit more producers of the same type. Most of these sellers also sell in other non-organic weekly markets closer to their farms, or through box schemes.

Another more recent social movement engaged in alternative food chains is AMAP (Association pour le Maintien d'une Agriculture Paysanne), which is very active in the Midi-Pyrénées. It is a sort of box scheme or community-supported agriculture project organised by consumers interested in small-scale sustainable agriculture. Their objective is to protect or increase the incomes of farmers by cutting out middlemen, a political agenda that is attracting increasing support in France. The concern with organic healthy food has reached the public mainstream, with documentary films denouncing the danger of conventional agriculture and the effects of corporate retailing on health. These include *Nos enfants nous accuseront* (2008), *Solutions locales pour un désordre global* (2010), and for the survival of small farmers, *Tous comptes fait* (2008). In addition, several councils have taken measures to promote organic agriculture on communal land. Local AMAP members commit to buying food for a year (or six months) in advance, even when farmers are not able to deliver, for example as a consequence of bad harvests. Unlike other outlets that are organised by farmers and rely on a historical tradition of exchange between the farm and the market, AMAP has an ethical and social dimension of solidarity (*solidaire*) agriculture and is supposedly organised by buyers. AMAP chooses farmers who are not necessarily organic but who are committed to producing in a way that respects people, animals and the environment. AMAP's principles expand on the Conf's charter, emphasising transparency in the purchasing, production, processing and sale of agricultural produce, and the geographical propinquity of producers and consumers.

Who Are the Organic Producers and Consumers?

Most organic producers are small, family-based farmers, working on marginal and often poor land, away from the rich plains. They work farms of between 2 and 30 hectares, with an average of around 15 hectares, and rarely employ paid labour. For most it is their sole source of income. Often this means living a marginal kind of existence: in my sample, three young couples did not own the land they worked and lived in a caravan or a yurt. Remy, 55, remarks that after more than a decade as a farmer he and his wife have only recently earned €900 a month. This level of income is common among the Tarn's small-scale organic and conventional farmers, who often live on the minimum wage (*salaire minimum de croissance*). The farmers produce fruit and vegetables, livestock and cereals, wine and artisanal products such as cheese and bread. Many of these producers are politicised: they are, or have been, involved in trade unionism, professional organisations or cultural associations, they often belong to ATTAC, have participated in civil disobedience workshops, and claim an anarchist (*libertaire*) position.

In the south of France as elsewhere, mainstream food retailing, with the option of a one-stop shop and the lure of bargain prices, has taken the majority of the population away from traditional weekly markets. However, supermarkets have not totally destroyed the local attachment to certain food practices. Gifting produce from the garden and homemade food to family and friends is still common. Families still socialise at lengthy meals, which may start at noon and linger well into the evening. The older generation continue to make produce that has been labelled as *terroir*: their own jam, spirits (such as walnut wine or sloe gin), *paté, foies gras*, duck in a *toupine* (a ceramic preserve jar), *gras-doubles* (tripe), snail dishes and local *patisseries*. They also grow their own vegetables, go in search of mushrooms, dandelions leaves for salads, and wild leeks growing in vineyards. Working people still shop at the Saturday weekly market for fruit, vegetables, meat and fish, and will queue a long time in front of their favourite cheese seller. They search for specific tastes and specialities. For example, those who still make duck in a *toupine* buy the birds from producers who kill them in the Occitan way, by bleeding, which guarantees the flesh will not rot.

In the context of food cultures and the search for taste, buyers of organic products may have health or ecological concerns, but they also seek quality – that is, food produced locally by people they know, even if they do not necessarily visit their farm. Although for the most part the consumers are middle class (entrepreneurs, doctors, engineers), there are also many retired people (from the teaching and medical sector) as well as the unemployed and people who cannot afford the cost of renting housing in towns. A small minority has marked social and political convictions; they position themselves as supporters of small organic farmers and refuse to go to supermarkets.

AMAP members' political commitment to local provisioning and short supply chains is accompanied by a return to cooking simple food in the non-commoditised and meaningful space of the kitchen. They exchange and seek recipes to cook newly discovered products, or to experiment in cooking known seasonal products. In some ways this approach to food quality is in contrast to the elitist, hierarchical systems of recognition which have been developed through complex and expensive regulatory organisations, both for organic and place-of-origin labelling (Pouzenc & Pilleboue 2009).

The Practice of Organic Farming and Its Values

Having established the background to the Tarn, with its farmers, organisations and food culture, the remainder of this chapter documents why farmers in this region choose to go organic and what this means to them. The interviews and the material focuses particularly on those farmers who refuse to sell to supermarkets. The chapter argues that though many of them were already politicised, their motivation to go organic is to be found in their response to practices in conventional farming and in their struggle to survive and maintain their autonomy.

When asking them about why they are organic farmers, the phrase which keeps recurring in the interviews is the search for autonomy. They are seeking to control how they run their farms, who they buy from and who they sell to. They seek autonomy in relation to the state, to the Common Agricultural Policy (CAP), and in choosing what constitutes relevant knowledge in farming practice. These strategic choices are regularly contrasted with conventional farming practices, involving 'productivism' and specialisation, which inevitably makes farmers dependent on powerful economic circuits (suppliers or supermarkets) over which they have no control. Their comments on autonomy often bring in related fields, such as the value of thrift, or ecological practices which are self-sustaining. Some also talk about autarky, invoking an older model in which the farm provided all the household's needs. The contrast between autonomy and autarky echoes the distinction between self-sufficiency and subsistence made in Chapter 2.

Jean-Pierre, 70, says: 'Through organics we can stay autonomous, as free as possible. More than the fear of some chemical molecules, it is the desire for autonomy. I have seen so many farms becoming more and more dependent on inputs'. Using natural and organic inputs rather than expensive pesticides and artificial fertilisers allows farmers to minimise costs, but it also reduces the risks attached to depending on an external supplier. One of the reasons Patricia, 45, stopped supplying sheep to a local (conventional) cooperative was that in exchange for a secure income the cooperative insisted she feed her lambs with their fodder.

In Jean-Pierre's view, 'the main motivation is to keep the skills of farming alive'. By this he means going back to the basic principles of good husbandry: small-scale subsistence farming, balancing livestock and field crops, and not harming people, animals or the environment. For Jean-Pierre, the old methods of farming that his grandfather used to practise correspond to the autarkic ideal:

> Small farms are infinitely more productive than big ones. Such farms have almost completely disappeared in France. Autarky comes from diversified production: a few animals, a duck pond, a bit of vineyard, some cereals… [T]oday we see farms specialising in monocrops or intensive animal breeding, which are both as horrific as each other, and are a risk to the environment. The CAP was a terrible mistake because there are 30,000 peasants who disappear every year because of it. It is a big agronomic failure.

The younger generations of farmers talk about autonomy rather than autarky. For Sébastien, 35, autarky means being self-sufficient in terms of production, and this is not realistically achievable: 'Our concern is controlling the destination of a product. It is not about living in autarky, producing everything yourself, but it is about trusting who we buy from'. The limits and limitations of autonomy require negotiation. When it comes to seeds, vegetable growers Martine and Sébastien do not take the risk of reusing their own; instead they choose 'small and ethical' suppliers: 'It is all about not depending on a system that kills you from behind'. On the other hand, it is in the name of autonomy that Martine and Sébastien refuse to sell to AMAPs, which are elitist in their eyes, and can sometimes also be paternalistic in dictating what farmers have to produce.

Two traditional farmers, who have used their old (organic) methods for the last 40 years, did try expensive synthetic inputs for a short while, despite their mistrust of modern products and technical advisers. The lack of efficacy of the chemicals and poverty were factors in them ultimately rejecting 'modernisation', but another important motivation was their attachment to peasant culture. The peasant ethos involves generating an independent income by using self-created and self-managed resources, rather than by borrowing money. Indeed, organic farmers often note that they may be earning less, but they are less in debt than conventional farmers who buy the latest technology. The Roucaries note that letting their guineafowl loose in the field, as their grandparents did, provided a great natural insecticide. The Py family refused to adopt the financial strategies of other (conventional) local farmers, who took on loans to produce pink garlic intensively, using up to 18 different treatments, though they did have to work harder as a result. Jean-Pierre Py, 40, remarked, 'in the time I spend weeding my garlic field, the [conventional farmers] have time to go to the beach [100 km away]'. Mrs Py, who is 85, interprets the move towards productivism in the 1960s and 1970s as a result of a desire to accumulate money. By contrast, they were more concerned with the values attached to the continuity of the farm, its productive bases (cows, land, seeds). She recollects:

There was only this room when I arrived here. After 30 years of marriage we managed to buy the 11 hectare smallholding next door. We were the destitute of the area, with old cars, real tramps. But we were putting money aside, and after we bought it, we were able to build the kitchen and to embellish the house.

For these traditional farmers, borrowing money from the bank and becoming indebted should be avoided at all cost, as it undermines the resource base and hence one's autonomy.

Among those who farm organically after practising industrial agriculture, it is common to find that the shift follows a period of reflection about the incoherence and contradictions of 'productivist' practices as advocated by the CAP. Daniel realised that he was only a 'node in a commercial system' following a series of events. A crucial one was a truck strike, when they could not deliver their 720 lambs to the cooperative. 'In the meantime, we had to feed them. It was very costly and in the end we had to sell the lambs to the abattoir. By then they were too fat, so they paid us very little'. There is also unease about

the way intensive agriculture affects the welfare of animals and leads to
the commoditisation of the 'living'. Daniel explains:

> You get to the point where you think: I am fed up seeing my lambs
> represented merely as bank notes. On the one hand, you are in a
> world of finance, and on the other, you live and work with 'living
> matter'. At night I had nightmares where I could see bank notes
> coming out of the arses of my ewes. Do you realise the contradiction?
> What is a 10,000 franc note? You can set light to it, it is not worth
> anything! And you have a living animal, who goes through the effort
> of birth, and this is worth a mere 10,000? So I was torn apart. It is
> because of this that we ended up going organic, with less animals
> and working differently.

The old practice of exchanging seeds is still widespread and
supported by organisations such as N&P, who promote traditional
varieties of crops that withstand drought and respond well to poor
soil and low inputs. Daniel grows old varieties of wheat and has
introduced local and robust breeds of sheep into his flock. In 2010 he
and his family went to the Middle Atlas Mountains in Morocco to meet
farmers, and brought back several kilos of seeds that he then planted in
his own fields. Michel grows an old type of local pea that has a square
shape, and sorghum from Africa. Unlike conventional farmers, who
prefer short plants, he grows an old variety of wheat with a long stalk
that gives him a lot of straw.

The mixed organic agriculture that many organic farmers practise
corresponds to the type of farming that was found in the area before and
just after the Second World War. To this older ethos of saving money
and energy, and avoiding waste, new ideas have been added, such
as ecology. But the decision to farm in this way is not a conservative
one, as it goes along with considerable invention and dynamism in
developing resources. As Michel 53, explains:

> We ask ourselves about the problems of the environment on a
> daily basis. Apart from the fuel for the tractor and the salt for the
> cows, I buy nothing and I produce everything. In terms of energy,
> it is through my cows and crops that I fertilise the ground, without
> having to buy any organic fertiliser. Everything is interconnected.

I work a complete system. I produce the straw for my cows, these provide the manure, which I put on the land, and I grow legumes and alfalfa to fix the nitrogen.

Furthermore, he set up a system to collect rainwater from the roofs of his premises in several wells. With a mobile solar pump, he takes this water to his cows wherever they are. Recycling has always been part of the peasant mode of farming. In order to recycle the whey generated by cheese-making and avoid investing in a costly system to treat pollution, Léonore, 50, chose to breed pigs that can feed on whey. The Py family produce their own tools from recycled machinery; they made a machine to hoe their garlic from an old road sweeper. This search for ecological farming methods extends to their own homes, where dry toilets are very common among new generations of *néo-rurals*.

Nathanael, 32, notes that being economic or thrifty and being ecological have a lot in common. Philippe and Edwidge (50 and 40 respectively) think that it was their poverty that first led them to use organic methods; they could not afford to buy feed for their chickens but had plenty of space for them to find food. This had a positive outcome, as local farmers and consumers came to appreciate that their baby chicks and goslings had a better life expectancy than those bought from conventional farmers.

A strategy of autonomy generates creativity in the development and efficient use of resources, both in terms of material things and knowledge. This creativity becomes a personal achievement, gained through reflexivity, entrepreneurial endeavour, hard work and moral principles. When he was rearing rabbits, Philippe found out how to tan their skins in order to make blankets for his children. After watching a TV programme describing the use of sprouted grain for feeding animals in Russia, he and his partner Edwidge implemented this idea. In their rustic kitchen, sipping herbal tea, I was shown two large basins with sprouted grains behind us. 'You need half the amount of grain, and it is richer in protein. Your rabbit's skin becomes lustrous and hens lay more eggs'. They are proud to point out that a neighbour switched to the use of sprouted grain for animal feed, despite the disapproval of her son, who had been educated in a farming college.

This is one of many examples where we find an explicit contrast between the innovative livelihood strategies of *néo-rurals* seeking autonomy, and those of conventional farmers with a more rigid view of the balance sheet and the desire to increase income. This reappropriation of peasant values by *néo-rural* farmers often takes on anarchist and anti-capitalist ideological tints that are not necessarily present among producers from a traditional farming background. Although it is difficult to generalise as some organic farmers stress ecological as much as social values, most of them defend a way of producing, working and living that puts profit as a low priority. 'Time is *not* money', says Philippe. Many pointed out, like Pierre: 'If I wanted to earn money, I would not have chosen farming'.

Earning money may be accepted as a necessary part of realising a livelihood, sometimes reluctantly as we shall see, but it is not an end in itself. Instead these *néo-rurals* prioritise other values, including the creativity of work practices, craftsmanship, the preservation of the environment and the quality of food. They also value cooperation and the creation of alternative social networks, as will become clear in the next section.

Several describe work as a source of personal satisfaction at many levels: the daily relationship with nature, the match between their ethical principles and their farming practices, and the freedom of being one's own boss. Jeremie, 28, who is just starting out as a farmer, explains:

We never really have much time, we work like madmen. It is hard when you have to get up and it is freezing cold, but you get up. Wealth is elsewhere. It is knowing that I manage myself, I don't have a boss, somebody to tell me how I should do things, that's where happiness lies, in my experiments, in the understanding of my soil, how things function around me. People tell me it is hard work. It is not. Working in an office, that's hard. I have worked as a youth social worker and I have always wanted to work with something that you can believe in. I hope I can one day assimilate both: get urban youth to come and work in the fields. I want to share. When I think about all the people who are somewhere they don't want to be! The expensive TV, the camping car, the home cinema, it is a flight. Here you cannot flee. If you are a peasant, you cannot flee. If tomorrow there is a problem with the cows, and even if you have thousands of other things to do, you have to deal with them now. When I see my salads, it is a pride – not in my skill, it is a nice variety of salad, with a really good taste – it is a pride to give pleasure to people through food. It is complicated to understand how and why.

Jeremie contrasts the un-rooted people of the West with those of Africa where he travelled and worked:

There, almost everybody is a peasant, everybody cultivates a field, or if not them their son or a relative. So there is a consciousness about nature, which is real and collective. Pierre Rabhi is so right to say that we [Westerners] are hydroponic children.

The quest for a quality of life and the search for meaning and coherence explains why Michel and his wife Annie gave up their very successful (conventional) goat-cheese business. Unlike traditional farmers who inherit land and knowledge (Salmona 1994), néo-rural farmers are more likely to reflect on why they have chosen the farming life. They felt they were working too many hours, and because of the large size of their herd they had lost the closeness to animals that they had initially sought. In the meantime, Michel had met an organic farmer and learned that it was possible to produce organically and live from it. His wife is an excellent cook, so they decided that she would make *patisseries*, using their own wheat flour and their own fruit and vegetables. They went organic, diversifying into wheat, leguminous crops and beef cattle. Now his wife works part-time and produces artwork in her spare time. Many farmers have taken similar measures to reduce working hours. Léonore, who would like to find the time to learn weaving, insists that, 'What is important too, is to have some time for oneself, one's family. Not always thinking about the bank account and work'. In order to gain more time, she milks her cows only once a day instead of twice, and this despite the fact that she uses a 'rustic' breed of cow that is less productive than those used in conventional farming. 'We have less milk, but it is a choice. We are not trying to break the record in quantity of milk or to have competition cows, but gosh, do they have a lot of cream!'

Organic farmers often state that what they supply is quality, whereas supermarkets provide quantity. For Pierre, 45, quality is more important than profit. He refuses to follow the advice of one of his organic colleagues, who grows bulky varieties of lettuce:

He is 100 per cent organic, that's not the problem, but his salads are less tasty. Because they are big, people feel it is good value for money, but they are less tasty. So he tells me, and maybe he is right, 'You will never make enough money ... If you want to earn a living, you need to produce bigger quantities.

Farmers think that increasing production usually means more time spent on financial and marketing activity and less on farm work. They also think that it means losing control over the quality of the product. So, if Lucien made more ewe cheese than his current 60 litres of milk a day allows, the quality would drop: 'When you have more volume, you have to be quick. You have to turn some cheeses every 15 minutes, and so if you make more, you can't turn them, so you won't have the same quality'.

As Jordy, 36, puts it, anything exponential is 'pathogenic'. A recurring concern among small organic farmers is the question of the size of the production unit. How do you define small or large? Is it connected to revenue, size of the farm, with the fact that they employ wage labour, or with whether they sell to retailers or for export? What is at stake here is their desire to distinguish themselves from intensive, large-scale, commercial organic production, which they believe cannot practise real organic agriculture. This is an important political agenda for N&P and other small organic farmers at a time when the new European organic certification is replacing national schemes, and which favours large-scale production for supermarket chains, with less transparency about farming practices.

More radically, Remy's refusal to be subject to the 'tyranny of the market' led him and his partner Catherine to forego commercial opportunities, and like some others in the Tarn made it their goal to produce 'non-elitist' wines: 'I make wine for people here, with salaries here. I make a good and medicinal wine but not too expensive'. He has also reduced the size of his vineyard. Remy argues that 'organics is an economic system. I do not want to live otherwise. It is a choice, that you can call de-growth'. The idea of 'de-growth' (*décroissance*) puts humanist and ecological agendas in opposition to economic expansion, and advocates scaling down production and consumption. Applied to farming practices, de-growth implies restrictive practices: growing vegetables in their proper seasons, not using heated greenhouses, and not driving long distances to sell products.

Reinventing Social Relations: Exchanging Labour and Goods

The previous section focused on farmers as they struggle for autonomy in production by positioning themselves in a particular way to factors such as time and money. A distinct but related set of concerns has to do with human relations. Sociability and cooperation are strong values. The farmers pursue social values in production, exchange and consumption, but sociability blurs the boundaries between these different moments. As we shall see, this occurs in labour exchanges, commercialisation activities, sharing knowledge, or just spending time together.

Workload and the availability of labour, both on the land and in small scale marketing, is a major issue in the survival of the farming household. Most have small families and need to find help through various networks. When vine disease hit the Gaillac region in the 1990s, collective building work groups (*chantier*) were constituted around a group of N&P winemakers. This tradition of cooperation in N&P, but also in the Conf, allows them to organise large work groups within a short period of time. In the present case, they took on the task of uprooting the vines and burning them before replanting. This form of cooperation is very similar to that practised by farmers until the 1960s, such as labour exchanges at harvest time, which always ended in a feast. The practice ceased with the introduction of combine harvesters, which led to a more individual mode of farming.

For Jordy, collective work groups (*chantiers collectives*), volunteering, woofing (international voluntary work on organic farms) and training placements all contribute to constructing new 'social forms' while allowing for the reproduction of the farm:

> The important task in reviving this type of agriculture is to create a social milieu and a culture that goes with it. Around Gaillac, there

is a good network that was built some 20 to 30 years ago. A group of friends united to make organic farming viable. It is this network that structured the commercialisation networks such as the yearly Biocybele fairs and the *noctambio* markets. Mobilising 100 people in a purely commercial system to popularise organics… that's why the social network around Gaillac exists. Because it has been 'cultivated' in the noble sense of the term. Strong networks between peasants and non-peasants, that's what I call the social network. We need such a network to get the farms running because it takes a bigger workforce to pick grapes manually than with a machine. You both have nice meals together and help in replanting, it has no [equal] value. That's the cultural side of agriculture. And it always starts with food, the pleasure of eating well.

Sociability also extends into developing markets. Léonore recalls the time when her group of friends decided to start a collective project selling both organic and non-organic products of *agriculture paysanne* (produced following the Conf's charter). They had already sold food together at political events, such as the May Day holiday, at pacifist festivals, and even in front of the prison where José Bové was imprisoned after the Macdonald's episode (mentioned above). In keeping with their ideology, the project had to involve small farms, which meant that they needed several providers to have enough stock to sell. It took some time to develop, progressing from a 'shop' in a van to a market at the farm, and eventually to a producers' shop in town:

We kept meeting. In fact it was one big meal after another. The ADEFPAT [a Midi-Pyrénées association for the development of employment in rural environments] woman got fed up and told us off: 'You just want to have fun together, you don't really want to earn any money'.

Jordy, meanwhile, argues for a particular view of exchange, as cooperation and gift-giving as opposed to self-interest and profit-seeking. This is part of a project to think and act in a way that can accommodate alternative human exchange relations (including between producers and consumers):

The only way to get out of this calculative thought that is polluting our minds is to trust in the collective. In fact, it is a way of relating to life. Life is generous. I give you something, and if it is not you who returns it, it will be somebody else. This becomes a group that experiments … in trusting the group, oneself, and by extension the life one is building. You have to unhook economic thought and cultivate the social network.

Following their vision has entailed constructing ecological housing, growing their own food, creating cooperatives of small artisans, and even an attempt at schooling their children outside the state education system.

These producers have developed a variety of exchange relationships in a more closed economy to provision themselves and obtain the inputs necessary to farming and living without spending money. In some cases, the relationships are valued not just because they reduce engagement with capitalist circuits, but because they explicitly avoid any kind of calculative logic. The gift of surplus food to avoid waste has always existed in local culture: garden surpluses are often given to neighbours and friends. In addition, they practise *troc*, a word used for a variety of non-monetary exchanges, including barter.

Non-monetary exchange is very common among N&P producers as it allows them to minimise their involvement in formal markets. The process often starts with giving away unsold produce to another stallholder at the end of the market – for example, a crate of tomatoes. Then the other producer returns the gift, for example a chicken. Léonore likes it on the grounds that it is an egalitarian social relationship, an alternative way of exchanging which creates links between people and sometimes produces dense social networks. '*Troc* is something I am very keen on. If it was up to me, I would use it all the time'. She started this form of exchange when she was younger: after having cleaned a filthy chicken shed, she was offered money or a kid goat. She chose the goat and later started making cheese. When asked how they assess what to give and in what quantity, she replied: 'That does not matter, we are not like the SEL [*systèmes d'echange locaux*, the French variant of what is known in Britain as a local exchange trading scheme, or LETS]. I give and I receive something. It is like if somebody was giving a gift'. When questioned about whether people take advantage, her 30-year-old son answered: 'There cannot be any profiteers because there is no demand, it is only an offer; we offer something to somebody, we do not ask for something in return'.

Troc furthers the strategy of not giving money away to the 'system' (variously 'capitalist', 'monetary', 'the supermarkets'). Remy, who is interested in local money schemes, has issues about the circulation of money that supermarkets keep as profit, as that feeds the 'system' rather then returns it to the community. Jordy explains how through the creation of a community bulk-purchasing project, not only did the group manage to save 15 to 20 per cent on the cost, but the scheme became the platform through which he and his partner now sell their own produce and organise cultural events.

Livelihoods, Markets and Customers

These farmers gain most of their livelihood from selling food, whether directly on the farm, through box schemes or at farmers' markets. That means they have to set prices, and this is problematic. However it is done, it brings to the surface the difficult relationship between their farming values and the wider commercial world in which they and their customers exist. In practice, most set their prices using market mechanisms, getting information from supermarkets and other outlets. The alternative would be to set their prices, and hence their income levels, from working out their costs of production, but that is not straightforward, either in terms of knowledge, or in terms of what constitute 'costs'. Two short examples give a flavour of the issues.

For Daniel and Blandine, who practise mixed agriculture and breed cattle and lambs, assessing the cost of production is not easy. One needs to calculate the CAP subsidies for each item, taking into consideration the fact that not all the grain is turned into flour because some is used as feed for their livestock. Among their outgoings are the cost of transport (to markets and the abattoir), the cost of the abattoir and butchering, and the cost of certification. Blandine remembers when their accountant produced figures for them:

It was a disaster: we had always been under the belief that the subsidies allowed us to make a decent return, when in fact they only covered half our costs. We lose money on all our products, except maybe for maize and buckwheat, where we break even. For example, we sell a kilo of lamb at 13.50 euros but without the subsidies we would have to sell it at 21 euros or more. For the wheat, we should sell it at 20 cents more per kilo, which would be too much for the buyer.

Labour is never taken into consideration as a cost. As Sebastien and Martine explain:

We don't fix the price in relation to the cost of inputs (seeds, manure or work), the hours spent weeding or cleaning carrots. We do it as a function of what's going on in the market, but there is a threshold that we cannot reach. For example, I sell my lettuces at one euro if they are small by weight, but if it is larger, I still sell them at one euro. We try to have reasonable prices that valorise the labour, but sometimes they are voluntarily low.

This brings us to the complex relationship with customers, a relationship essential to the farmers' livelihoods, but also sometimes seen as a source of dependency. All these producers sell direct (*vente directe*), a practice that was until recently scorned by conventional farmers as old fashioned. The practice of direct selling is something they had to undertake, even if they report slow growth and initial reluctance, because, as Jordy explains, it is the only way to resist large retailers:

Local trades need to help each other, to avoid having to deal with intermediaries, so as not to be the victim of the commoditisation of things. We need ties and social meaning, that's the key word; to be in a relation with others, in permanent human contact, so trade does not prevail over other values.

Many of these farmers try to educate the rootless 'hydroponic' consumers about their role in supporting them through buying direct, but also about the seasonality of products, the difficulty of farming organically, the absurdities of state regulations, and the value of their labour in producing quality food. This is why most of them emphasise the importance of face-to-face contact with their customers.

Farmers have learnt the significance of communicating with their customers and informing them about farming. One method is via a newsletter providing information about progress on the farm, recipes using unfamiliar ingredients (buckwheat, spelt), information on bakers using their products, or on different techniques of milling grain and its effect on the nutritional quality of flour. It is common practice among N&P farmers to organise a yearly visit to their farms, and other events which always involve sharing food and information. The effect of the visits is to increase transparency about farming practices. Producers are sometimes reliant on the labour of their consumers in farming activities (weeding, fruit harvesting and grape picking), but also in supporting them in legal battles, for example creating a support group against vaccination in the case of bluetongue (a non-contagious insect-born disease that affects ruminants) which was found to be inefficient and dangerous for livestock.

Remy challenges the binary use of the terms consumer and producer on the grounds that it is a false dichotomy: 'We are all consumers and producers'. The concern to erase this opposition is rooted in a belief in horizontal and egalitarian relationships that is part of the history of N&P, which at an early stage relied on cooperation between farmers and their customers. Like the Conf activists described by Williams (2008a: 112), they create and cultivate relations of solidarity that pursue the logic of sociality rather than profit, and collapse the distinction between consumer and producer. A minority of faithful consumers are politically committed to this agenda, usually those involved in AMAPs. For example, in Gaillac's small organic market, some consumers took the initiative of holding a stall selling vegetables in winter, which they gather from several producers as these would not come themselves to sell just a few items.

Other farmers, however, deplore the fact that some customers do not question the agro-industrial food system and the state policies which are so detrimental to small farmers. N&P has created an alternative, non-state, non-official certification scheme which sets out purer and more radical organic practices. There is not necessarily an outright condemnation of consumers, possibly because they are dependent on them for survival, and also because some producers are aware of inconsistencies in their own attitudes and those of their children. They also note that many consumers are not aware of the ramifications of cheap food policies, even if they pay for them indirectly through taxes, as these are used to generate subsidies.

Daniel is one of the rare farmers to express frustration over the power of the consumer, who ultimately decides to buy or not. This 'power of the consumer who holds the wallet' does not necessarily come with a sense of responsibility and personal involvement. He argues that, after questioning the mainstream food system, AMAP consumers should not delegate so readily to farmers the responsibility for auditing practices. Daniel has an idealistic view of his relationship with consumers, which he views as a collaboration in creating an alternative economy. His job is to provide them with produce; theirs is to be political activists: 'We need the citizens to take over from us at that level because we cannot be at the demonstrations and in the fields'.

Conclusion

Although there is a minority of farmers who chose to produce organically for a niche market, and who sell to whomever they can (supermarkets, organic shops and even export markets), many of the Tarn organic farmers, despite their diversity, share a common political concern centred around autonomy vis-à-vis these markets and state policies (including the CAP), as well as holding to a range of social values which they oppose to monetary values: reciprocity, cooperation, solidarity, friendship and equality. They are concerned to build a sustainable agriculture for future generations. They think that exchange should not be 'calculated' and requires generalised reciprocity.

In order to practise their version of organic sustainable agriculture and keep their autonomy, they use various strategies, including self-provisioning in food through *troc*, exchange of labour and inputs, cooperation in use of materials and knowledge, selling direct and educating consumers. These strategies draw on local, traditional, farming practices, both technically and socially. But they also demonstrate 'social creativity' (Salmona 1994: 104) in the constant search for new solutions and the belief in experimentation and reflection that goes with seeking information from others. Indeed they have often been avant-garde, developing techniques and methods that were then followed by mainstream agriculture (for example, minimum tillage systems). They generally seek to earn a living without sacrificing the quality of their life, that is by finding a balance between work and leisure. The consumption and exchange of good quality local food is central to that goal.

To some, going organic is a lifestyle choice that goes beyond autonomy, and beyond the ethical and political decision not to contribute further to the degradation of the environment and to the social inequalities inherent in the agro-food system. They embrace a utopian social project of building an economy or community outside commercial mainstream society, one based on new forms of more egalitarian exchange between people, and an alternative local food supply chain outside that of the large retailers. Yet, they have become dependent on the buying power of consumers, some of whom act and think in conventional, hedonistic and commercial ways, at odds with the farmers' utopian and purist project. In addition, they often find it difficult to reconcile their desire to keep prices low with the cost of production and commercialisation.

Finally, another danger for these farmers, who are engaged in a political project of purification and of autonomy from the market and mainstream consumer society, is the risk of self-exclusion from society, which may imply that they are less likely to engage in and transform it. Williams (2008b) discusses the risks of self-exclusion in the case of the Conf activists of Larzac. On the other hand, farmers are different from activists in that they produce food for a society which has lost awareness of its dependence on them (for eating). In addition, it is in the daily battle with the natural environment and against state policies and regulations that these organic farmers became radically politicised. Yet, they actively break this marginality and isolation by meeting their consumers, and they strive to apply their principles in their daily practice. Their attitude may indeed attract future generations in search of meaning, moral principles and justice. A sign of this is the fact that despite the progression of the 'conventionalisation' process, there are more and more small farmers who choose to go organic in the Tarn and Midi-Pyrénées. Moreover, there are still not enough of them to satisfy the increasing number of consumers of organic produce, both locally and in the more populous coastal region.

Sources

This chapter is based on data collected during fieldwork in the Tarn, where I have close family ties. More than 30 organic farmers and food producers were interviewed, as well as consumers; interviews were carried out at markets and in people's homes and workplaces. I would like to thank all the organic producers and farmers who helped with this research – including many who are not mentioned in this chapter – for their willingness to meet me despite their busy schedules, and for talking so readily about their values and strategies, joys and struggles. I would also like to thank my mother, a keen supporter of local and sustainable farming. Without her the research would not have been possible. Special thanks also go to Marie-Claude Bousquet, who introduced me to more traditional organic farmers.

The main literature on *néo-rurals* by sociologists Bernard Hervieu and Daniele Leger-Hervieu is based on ethnographic research done in the Cevennes, Hautes-Alpes and Pyrénées regions between 1976 and 1980 (see Leger & Hervieu 1979; Leger-Hervieu & Hervieu 1983). In the earlier of these two works, they argued that the *néo-rurals* belong to the new petty bourgeoisie and have abandoned their original anti-capitalist philosophy in their endeavour to secure integration into the local society and economy, even if they kept an anti-urban, anti-modern, anti-institutional romantic ideology. The second book deals with the religious dimension of this utopian movement. Fournier (2008) provides a good overview in English of the extensive French literature about 'de-growth'. Information on Pierre Rabhi is limited in English to Wikipedia and Baykan (2007). The sociologist Deléage (2004) focuses on anti-productivist farmers. Leroux (2011), a PhD on organic agriculture in the Midi-Pyrénées region, focuses on mainstream organic producers, whereas Van Dam et al. (2009) consider the conversion of conventional farmers to organic agriculture in Belgium and the north of France. The journalist Philippe Baqué (2012) has edited a book that critiques intensive organic farming and offers information on alternative approaches: it includes data on the main political and commercial actors in the French organic landscape, on AMAP, agro-ecology, Via Campesina and the right to the land (see also www.reclaimthefields.org.uk). Finally, a recent book in French by Quellier (2007) on the history of food considers the relationship between the countryside and cities, and emphasises the economic and social significance of vegetable plots for both urban and rural dwellers.

6

Andalusia, Spain

Pete Luetchford

Our third case study from southern Europe focuses on political cultures and strategies behind local and organic food provision in Cadiz province, western Andalusia. Here we find a variety of small-scale organic producers. Some are *campesinos* who run inherited family farms on the margins of the great estates, and come from farming traditions with a history of supplying local towns and villages with food. Others have roots in the political movements of agricultural labourers who occupied land; these produce for local markets, and through their connections supply food to major cities. This chapter picks up these different strands and explores the way they intersect over time. A central part of the story is a failed attempt to draw farmers with different backgrounds into an alliance based around a second-level retail and distribution cooperative called Pueblos Blancos.

Pueblos Blancos was conceived as an attempt to provide more stable livelihoods for smaller farmers growing organic fruit and vegetables for local and regional markets. Pooling production would, it was hoped, allow these farmers to supply a greater range and quantity of produce, and so consolidate organics for local and regional consumption. The case is instructive in the first instance because it allows a discussion of ideas and practices that have inspired people to try to build closure into economic relations. These ideas and the commitment to alternative economies are an important part of the story; they pre-date Pueblos Blancos and continue to drive people's strategies.

On the other hand, the case speaks to problems encountered in attempts to build alternative economies when people engage with open markets. For Pueblos Blancos there was the ever present danger of market competition from the larger farmers and commercial interests that dominate the organic sector in Andalusia, but which had hitherto confined themselves to export-oriented activities. Another challenge was to build viable networks when participants come from different political cultures. A third danger was that a more commercial organic sector would reaffirm the distinction between producers and consumers, so reproducing features of the open economy. Together, these questions about the viability of a project to build a regional system of organic provision in Andalusia speak to the problem of building and maintaining closed economic relations in the face of the open economy, even when there are long and enduring traditions of closure.

'Land for Those Who Work It'

My story of an alternative politics of food begins with the activities of the field workers' union, the Sindicato de Obreros del Campo (SOC) in western Andalusia. In the late 1970s and early 1980s, the SOC led rural workers in demands for 'land for those who work it', a radical programme of civil action for agrarian reform. The SOC lobbied government, invaded and occupied land, blocked roads and farm gates, and in the process clashed with the Civil Guard. The outcome of these actions was a number of cooperatively run farms. These cooperatives had common aims and ideas, but followed a range of trajectories and strategies. What they shared was an agenda that responded to the direction mainstream agriculture was taking.

A key feature of farming and hence political culture up until the 1970s had been the predominance of large tracts of land, *latifundias*, arranged around *cortijos* (farmsteads), which were owned by absentee landlords and run by an overseer employing temporary workers. Employment involved informal, verbal and temporary contracts, often on a daily basis for the harvests and other peak periods of work. These *jornaleros* lived in nucleated settlements or *pueblos*, and were hired in the central square and taken out to the land to work. The outcome was a system of flexible but uncertain employment, which suited employers but kept workers in conditions of near or actual starvation. The last major European famine occurred in Andalusia in 1905, and older people today still recall starving people roaming the countryside and stealing food. Under this system, employment and labour relations between landowners and workers was marked by the use of force and moral indignation, as codes of conduct were contested and violated (Martinez-Aler 1971). While liberals introduced various reforms to try to defuse labour problems, labourers and landlords tended to remain entrenched in their respective left- and right-wing political traditions. In any case, for labourers, influenced by the writings of Marx and the anarchist ideas of Bakunin (brought to Spain by the Italian activist Fanelli in 1868), politics meant struggle against class oppression.

All this began to change in the 1970s, a time of great change in Andalusian agriculture. Machinery and agro-chemicals entered the sector, so chronic unemployment threatened field workers' livelihoods. The coming of herbicides and machinery also meant increasing concentrations of land ownership and expanding monopolies. The larger farms today continue to practise extensive farming of cattle, olives, wheat cotton, sunflowers and, in some areas, wine. With the coming of mechanisation, the *jornaleros* were increasingly redundant; it seemed to signal their end as an economic force. Many of the million or more Andalusian *jornaleros* who had once worked the land abandoned agriculture. What remained was a political identity and tradition carried over into the towns and cities, and absorbed into new workplaces and forms of politics. Many migrated to the coast to work in mass tourism, to the cities to seek employment in industry, or to northern Europe to work temporarily in harvests or in factories. In Andalusia today, the many squares and streets named *de los emigrantes* are a reminder of the exodus. But as many as half a million *jornaleros* remain today welded to the identity (Donaire 2011), and most subsist on casual and temporary employment and state benefits. Others continue to organise through their unions, and draw on the tradition of radical politics of anarchism, socialist and Marxist-inspired ideology. A handful of these activists revived the practice of land occupation last seen in the 1930s to realise the long-held aspiration for land distribution, or *reparto*. In this endeavour, they welded their radical politics to a model of agriculture that had long existed alongside the great estates.

Between the *cortijos* there exist quite substantial but scattered pockets of smaller farms and kitchen gardens (*huertas*). From the former we get a tradition of market gardening by autonomous commercial small farmers. From the latter, come notions of self-sufficiency in food;

there are small family-owned *huertas* and municipal allotments on the periphery of most towns. The smaller landowners, squeezed by economies of scale, were natural allies for landless workers struggling to retain employment on larger farms using an increasingly industrialised agriculture. So, in appropriating land for those who work it, the SOC embraced the model of small-scale mixed cultivation, but in keeping with their politics they wanted, in the words of one activist from the time, 'to vindicate the idea that land should not be private property but in the hands of workers'.

The process through which labourers achieved that goal, their approaches to making livelihoods, and their consequent experiences have varied, but all responded to the dominant model of agriculture. Some, such as the large cooperative at Marinaleda, with 2700 workers, have pursued more industrial methods and a communist model. Others, such as REPLA, have concentrated on generating rural employment and produce in large enough quantities to export to Tesco. A third cooperative, Tierra y Libertad, operates a show farm, conference and activity centre. Of more concern here are the 'autonomist' and arguably more radical agendas pursued by El Indiano and La Verde. In the case of El Indiano, workers engaged in protracted confrontations, and gained ownership of land after many years of struggle, occupation, imprisonment and an eventual negotiated settlement (Romero 2003: 450–53). Once the scene of occupation by thousands of rural workers, El Indiano was also the venue for the anti-capitalist Second Zapatista World Forum against neoliberalism in 2003. Today the farm has an abandoned air, and a handful of workers keep goats and practice a mix of organic and conventional agriculture.

More instructive for the politics of alternative food provision is the experience of La Verde. Here, SOC members successfully petitioned for unused land owned by the state (Luetchford et al. 2010). The people at La Verde have spent years creating their version of organic production, which they call *ecológico*, and an alternative food marketing network. As the most successful and best known of the original SOC cooperatives in Cadiz province, La Verde has become a hub in an alternative organic food network. The importance of La Verde lies not in the size of the operation or its success in the market. Rather, it provides a basis to discuss the relation between politics and the cultures of food production, distribution and consumption, and the tensions between open and closed economies in southern Spain.

La Verde: A Politics for Production

La Verde began in 1986 when about 20 members of SOC from the town of Villamártin petitioned for ownership of 3 hectares of unused land alongside the Guadalete River, which was owned by the water board. In 1987 they were granted concession and began to grow food in earnest. Today, three remaining members of that group, Manuela, Manolo and Enrique, maintain cooperative membership with three more recent recruits, José, Isabel and Francisco. Over the years, the cooperative has expanded to 14 hectares, through further concessions of public land and the purchase of 5 hectares with money lent by urban consumers. On this land, and because they can irrigate it from the river and have a propitious climate, the cooperative are able to grow a wide range of fruit (plums, apricots, figs, blackberries, strawberries, grapes, nectarines) and equally diverse vegetables (onions, potatoes, leeks, cabbages, beetroot, asparagus, sweet potatoes, aubergines, peppers, tomatoes, lettuces).

The La Verde cooperative is well known in organic and environmental circles in Andalusia. Numerous technical studies and newspaper articles have highlighted their pioneering work in building a sustainable, closed, production system. Although by no means the first organic-certified farm in Andalusia, they are renowned for maintaining a radical agro-ecological stance, distinct from mainstream organics (Sevilla Guzmán and Alonso Mielgo 2005). At the level of production, this manifests itself in specific agricultural practices and social relations at work.

A fundamental aim is to eliminate relying on agro-industry. Organic standards permit a wide range of commercial products, but these are shunned because they would tie them to the interests of capitalist multinationals (see Luetchford et al. 2010). For example, visitors sometimes bring organic treatments as gifts, but they are not used. As Manolo points out, 'that container of Rotenone [a naturally occurring pesticide] has been lying around for years'. Another tactic to cut dependence on commercial suppliers is on-farm seed production and seed exchanges. The large seed bank is a resource to ensure genetic variety, reduce cash expenditure and resist genetically modified (GM) crops. The cooperative is an active participant in the anti-GM group, Red de Semillas. A third strategy is to minimise expenditure on machinery by repairing and recycling what they already have. Such practices help them create as far as possible an autonomous and closed agricultural system, outside mainstream commercial circuits.

The politics of refusal of commercial inputs is complemented by alternative cultivation practices. The only regular input is free manure, which they collect from local shepherds and goat-herders, and the livestock brought in to graze and so clear and fertilise the land. This requires reciprocal agreements with local shepherds because their refusal to inoculate livestock has created conflict with veterinary authorities; they retain their chickens but had to get rid of their pigs and sheep. The crops are grown in strict rotation to allow soils to regenerate and avoid the accumulation of pests and disease. This distinguishes them from farms, both organic and conventional, which practise some form of monoculture. Pests, like potato beetle, are controlled by hand, losses are tolerated, and insect habitats are created by tree planting, and by leaving 'weeds' around the crops. One of the achievements they flag is replanting the river banks with trees. The weeds and trees generate what they term 'ecological infrastructure'. An overall effect on the environment, they say, is to reduce average on-farm temperatures by 5 or 6 degrees centigrade. Farming practice based on agro-ecology therefore provides one set of distinctions they make between themselves and conventional and commercial organic farmers.

A second distinction rests on the use of intensive manual labour and the organisation of working relationships. The members attribute much of their success to their open-door policy towards people and new ideas. As Manuela says, 'we have never been closed; we have always been an open group, ready to learn from other people'. La Verde is an experimental station, and forms a hub for the exchange of information on organic production and seed varieties through talks, workshops and visits. Over the years, hundreds of visitors have come to stay, some for up to two years. The presence of these visitors indicates particular social relations of production. The volunteers work and learn on the farm through 'practice'. As Manolo said, 'I like books and reading a lot, but you only really learn through practice'. On the other hand, La Verde does not pay wages for labour, apart from to members of the cooperative. In the past they have employed workers but found it complicated:

> We began to understand that it was a really dangerous situation we were getting ourselves into. It led to problems working with non-members, not because they were not members, but because they were friends and family, and the profits began to fall. It required living in that hard economic world and having a business-like attitude, and that has always been a really difficult role for us.

Only using 'social labour' is grounds for a distinction between them and specialist producers:

> [Specialist producers use] an industrial style, with industrial machinery with workers, mostly Moroccans, who work under difficult conditions. They are able to put carrots on the market at €0.75, and we cannot sell them for less than a euro. And that is because we work in a more personal way, and the dynamic is different because we don't just grow carrots.

Work begins early at La Verde, especially in summer when temperatures exceed 40 degrees centigrade by mid afternoon. The main tasks involve planting, harvesting and weeding, using the short-handled hoe, or *soleta*. The tempo of work is relaxed, and there are frequent breaks as people stop to chat, wander off to start another task or take a stroll. Often work seems haphazard, piecemeal and ineffective. Certainly it is not geared towards maximising production, as large areas remain fallow at any one time. There is little tendency to 'self-exploit' to intensify production. On-farm social relations are lived through sharing tasks and sharing food. The workers are encouraged to take away any supplies of fruit and vegetables they want, and eat meals on the farm. By 9 AM it is time for breakfast, and people gather in the kitchen to eat bread *molletes* laced with olive oil and tomatoes, drink coffee and chat. At lunchtime, those who do not return to their homes in the *pueblo* eat together, using produce from the farm or procured from other producers in the network. One visitor who stayed several months, Antonio, pointed out that there are few places where you can wander outside, gather what is around and make a meal.

Manuela says one of the best things about the job is that there is no 'boss breathing down my neck, telling me what to do'. This telling statement demonstrates the objective of creating a collective of autonomous workers making a livelihood from the land. Personal autonomy is the aim: 'to free ourselves is a life project', says Enrique. Nevertheless, this commitment to individual freedom and free association raises difficulties. A key issue is retaining members and recruiting new ones. One way of explaining this is that the flip side of freedom is the requirement to control the cooperative and its trajectory. Only three of the original occupiers of the land remain, and they are conscious of the importance of the next generation, but the young are often not inclined to work in agriculture and, while the cooperative is open to temporary visitors, there are issues when it comes to converting people to full membership.

A second problem is subsidies. As a collective, La Verde made the difficult decision to become a cooperative. It was difficult because cooperatives were, in their minds at the time, hotbeds of corruption. Despite this, legal conversion into an Andalusian Cooperative Society (SCA) meant they could apply for subsidies:

> We had to decide whether we would go for 'public help', for subsidies, or not. As it was public money, we decided that if we didn't get it, it would go to someone else. The biggest landlord from my *pueblo* was getting 40 million pesetas of subsidies, though that didn't stop him shamelessly moralising about how undignified it was to live on handouts. There was no possibility of getting that public money if we were not a cooperative, it was the only way.

Subsidies may have been necessary to ensure the survival of the cooperative, but they play a contradictory role in compromising autonomy: 'people like my [Manolo's] father understood this. If you receive something for nothing, what morality have you left to ask for what? I mean, if they give it to you, you are left without your own will'.

As we have seen, La Verde has constructed a particular environment and a specific set of social relations organised around collective production, all built around a politics of autonomy. These practices are at once oppositional, in that they are a response to corporate industrialised agriculture, and alternative, as they try to work outside mainstream commercial circuits by relying on thrift and collective effort, rather than by throwing money at agriculture. This was well expressed by Enrique:

> If you arrive in a place, say 10 hectares, with 40 million pesetas, and set the thing up, it is sure to work. But if you have nothing and all you put in over the years is your energy, your hands, your imagination ... [T]hen that for us is sustainable agriculture, because of the tremendous effort you have put in.

The people at La Verde have been innovative, oppositional, and alternative in their production strategies. Ideas about freedom and personal control of one's destiny are no doubt a reaction to the historical experience of *jornaleros* as subject to the whim of landlords. In taking control of their own land and labour, they have been largely successful in escaping from this subjection. But at certain key points – the need for legal recognition as a cooperative to engage with the state and subsidies, the problem of recruitment, the use of diesel for machines, requirements that their livestock be vaccinated – their vision of an autonomous closed system is challenged as they are forced to engage with wider industries and authorities.

More problematic is engagement with open markets. In Chapter 2 we saw how a 'just price' is a perennial concern for petty-commodity producers who always seem to be at a disadvantage when they sell their goods (Luetchford 2008a). The way La Verde have dealt with this is by drawing on specific political and cultural traditions in their engagement with other producers and consumers. The collective began as a project to access food for consumption. Manuela says they occupied and petitioned for land 'because we had to produce to eat'. Once they had the land, their subsistence activities, geared towards reproduction in adverse circumstances, yielded a surplus. This allowed them to move into markets, and generate markets of their own. The process of how strategies and ideas about consumption and social reproduction were widened into circuits of exchange, and problems and tensions from engaging in the open economy, is the subject of the rest of the chapter.

Local Strategies for Consumption and Distribution

The members of La Verde agree they opted for organics for three reasons: they had to produce to eat, through SOC they were in contact with ecology groups, and locally they knew of cases of poisoning through pesticide abuse. We have already seen how their production strategies were a response to industrial agriculture. We now turn to their need for food and how they approached that necessity, before describing social and political connections with other farmers, activists, politically-motivated consumers and wider networks.

When they started to produce from the land at La Verde, the workers adopted a peasant model of economy. The key aim is to build as far as possible a closed system, avoiding external inputs and reproducing rather than expanding. Of course, there are expansive and even entrepreneurial aspects to their market interactions, but these are secondary to the original and primary aim: they 'had to produce to eat'. The background to this is, first, an experience of extreme poverty in a rural environment in which up to 80 per cent of the population of their town were day labourers whose precarious incomes were threatened by industrial agriculture. Hiring on an informal basis allowed landowners to exert pressure on political agitators and enforce a compliant workforce. Failure to be chosen for work could mean starvation. Manolo explained that exploitation of labour and maximising landlords' profits was a deliberate policy of the Franco regime (1936–1975). The profits were expropriated rather than reinvested, and the Spanish economy at the time stagnated. Enrique said that, 'for 40 years nothing happened; the economy just stood still'. In such circumstances, securing food outside the money economy gains great significance.

There were different avenues for increasing food security in precarious times. First, 'wild foods', such as snails, asparagus, and cactus fruit could be collected from the countryside. Enrique pointed out that 'the people here aged 50 and above are from a rural world. We were always in the countryside and we knew the things we could collect to eat'. By the 1980s that situation had changed, as rural employment declined and people became neo-urban. At the same time, democracy, EU membership and mass tourism were part of a growth in prosperity. As incomes rose, it no longer became necessary to scavenge food from the countryside. Nevertheless, such foods are still remarked upon, hold an important place in collective memory, and are gathered for home consumption and hawked on the street. To go to a bar and eat snails is a social occasion.

A second source of food is the *huerta*, or kitchen garden. Food production on these plots is a form of agriculture that has long been practised in Andalusia, and throughout Spain. The primary aim is the production of fruits and vegetables for home consumption. These can be supplemented with nuts and fruit from orchards, chickens, and cold cuts from pigs killed on the farms. With little cash to buy food, let alone industrial inputs, the workers at La Verde were drawn to traditional methods used by smallholders and gardeners to secure household food supplies. Manuela emphasises this idea: first and foremost, she says, 'our cooperative is a *huerta*, with products that we eat'. Following the *huerta* model meant growing food for consumption and struggling to be 'self-sufficient'. It was developed out of expediency, to take people through hard economic times.

The *huerta* model also gave La Verde a context for their agro-ecology; it provided an association with traditional practices for producing fresh, non-industrialised foods before the arrival of chemical inputs in the 1960s: 'We began working [the land] as a *huerta* for self-consumption. We worked it organically because we were going to eat it ourselves, we didn't want chemicals. We wanted healthy things to eat'. But this is a social as well as a practical model of economy. By following specific food-provision practices, La Verde draws on the idea of localism: a closed system based around the *pueblo* and the *huerta*, run by families to distribute food through personal and kin relations. In this way, people without land could hope to access food by drawing upon direct social relationships in their *pueblo*. This has two key aspects useful to La Verde: it gives a context and rationale for their production practices, and it affords an avenue for direct exchanges with local people who want to consume their products. In conversation, they often talk about the people who visit them to buy food. Indeed, at various times they have experimented with a farm shop: 'That [farm] shop was our most important outlet. Loads of people would come with their children in the afternoon to have a stroll and buy stuff'. On-farm sales were scaled down when they became involved in the second-level distribution and retail cooperative, Pueblos Blancos, in 2003. One outcome of this was that storage facilities and the distribution centre were moved to a warehouse with a cold store a few kilometres down the road.

Specific kinds of food produced using 'traditional' practices means an easy association between the *huerta*, fresh local produce and organics. There is no necessity that *huertas* should be cultivated without the use of industrial inputs, and some people today admit or are said to apply chemicals, sometimes in large quantities. But the fact that this is remarked upon, and that these gardens have a history that precedes the availability of agro-chemicals and manufactured fertilisers, makes that connection an easy one. The shift from local to fresh to organic products is traced in this statement from Manuela:

> Here in this *pueblo* there was a real tradition of *huertas* all along the riverbank. In the beginning, the people didn't buy our produce because it was organic, they bought it because the products were fresh and from here; cut in the morning and sold in the afternoon. But over time people began to understand that organic products were good for your health. That began with people like doctors and teachers. The ordinary people just bought it because it was fresh.

How readily people associate local food provision, non-industrial foods and organics was evident in interviews in one *pueblo* on consumption practices with 30 women of different ages and from a range of backgrounds. This data supplemented information collected at ten workshops that had been organised by local activists to promote organics to women's groups and parent–teacher associations in towns and villages around Cadiz province. The presentation on food additives and the organic alternative was followed by lively discussions. Women enjoyed describing the 'simple', traditional dishes and foodstuffs that had sustained their families over generations. The foods, redolent with taste and meaning, were repeatedly juxtaposed to modern 'junk', especially pizzas, which were associated with youth. The presentation on organics provoked conversations about their neighbour's eggs, the taste of plums from a particular tree, old-style breads, or the killing and processing of pigs on the farm. The discussions recalled life before food processes became industrialised, when every *pueblo* had a flour mill, an olive press and an abattoir.

The second feature of local food useful to La Verde was the importance of historically and socially embedded exchange relations outside mainstream commercial circuits. The majority of the women at the workshops spoke about the *huerta* they ran, or how they accessed home-grown produce through their extended family. Others paid neighbours for fruit and vegetables. Meat, oil, eggs, dairy products and bread have long been produced and bought through local sources. While people used supermarkets, most claimed to prefer locally-owned grocers' stores for their everyday shopping. On the other hand, many make weekly or monthly trips to larger towns to shop at chain stores. Such excursions are an opportunity to stock up on industrially produced bulk items, such as soap powder and canned goods, and to economise. They also provide occasions to shop for clothes, larger items and specialist products, and to enter a more cosmopolitan environment with urban and foreign shoppers. In a sense, these shopping trips have an air of being special events one dresses for, a treat for oneself and the family.

It is hard to gauge the importance of these different avenues to access food. Local exchanges certainly evoke commentary and have ideological importance. On the other hand, supermarkets are expanding and the tendency for local farmers to abandon the *pueblos* and sell into specialist organic shops in larger towns and cities suggests a slow decline in more traditional food routes.

Nevertheless, access to land puts the people at La Verde in a different category in the local food economy to that of *jornaleros*. They are able to produce for themselves and sell their surplus. Hence the *pueblo* became 'their earliest and most important market'. What is more, an economy based upon 'subsistence-plus' gives common ground with the small family farmers producing food for direct sale in the *pueblos*. The material expression of that common ground was the formation of Pueblos Blancos. This came about when Manolo from La Verde was employed as an advisor and extension officer on a project to develop organic agriculture run by the provincial council (*Mancomunidad*). Through this initiative, La Verde became involved in a second provincial council project aimed at marketing rather than production, named Pueblos Blancos after the famous white towns of the region. La Verde took the lead, consolidating a membership of local farmers and SOC cooperatives who practised organics or were open to transition and certification, and supplied their *pueblos* with food. In this way, La Verde became the lynchpin of the new organic retail initiative, as they bequeathed their network of contacts to the new cooperative. They also had a hand in employing personnel: Maria, who took the post of manager of Pueblos Blancos, was a personal contact, and they supplied the driver and people to work in the warehouse and offices.

Pueblos Blancos developed as an alliance between cooperatives of politically inspired former *jornaleros* and more individually motivated family farmers. The different agendas of these people necessitated a fragile truce that did not last, but the two types of members shared strategies and some ideas. Like La Verde, most of the family farmers in Pueblos Blancos have long supplied kin, friends and their *pueblo* through stalls in their town marketplace (*plaza de abastos*), or sell through local shops. The political culture is clear: there is a preference for local markets and consumers. La Verde insist that their first and top-priority customers are from their town and surrounding villages. The political culture was also emphasised in the Pueblos Blancos mission statement: 'Establishing short commercial circuits' was the objective; they wanted 'people who live nearby to benefit from eating their products'. Discussions with the members of Pueblos Blancos often turned to the possibility of opening shops and market stalls locally under the cooperative name, but practical constraints and more lucrative sales to customers in larger towns thwarted those ambitions.

Alongside the values intrinsic to forging local production–consumption links, there are economic arguments about competing in terms of money value by 'selling direct, without intermediaries'. During interviews, farmers in Pueblos Blancos insisted that they can compete on price with conventionally grown crops as and when they sell directly to the consumer. Whereas specialist production, national distribution chains and export markets require economies of scale with which they cannot compete, there are claims that there is scope to be competitive in other markets. For example, one founder member of Pueblos Blancos, Antonio Perez, produces asparagus, tomatoes and artichokes, and has a network of customers in his *pueblo*. He says he 'guarantees to be able to compete with any conventional producer on price as and when the sale is direct to the consumer'. The competitive edge is borne out by the success of Antonio Mulero's market stall in his *pueblo*. Antonio supplies a stream of customers drawn by his local reputation as a fruit and vegetable grower (*hortelano*), and knowledge that his products are grown down the road and are freshly harvested. Notably, his prices compare favourably with others in the same market and with supermarkets, and he plays little on his organic status; an inconspicuous and faded sheet of paper taped to the wall of his stall is all that signals organic certification.

Local production–consumption links therefore suggest both an alternative to capitalism – a rural world maintaining relationships in which producers and products are known and trusted – and opposition to capitalism, where these conditions are not thought to be satisfied. On one level, La Verde encapsulates that combination because it pursues a model of localism and self-consciously recuperates practices and traditions that 'go back thousands of years', such as water management systems learned from the Arabs when they occupied much of Spain. At the same time, La Verde opposes the industrialisation of agriculture in pursuit of profit, with all that entails for the rural environment, as well as the kinds of distribution networks that capitalism creates. This is clear in Manuela's account of selling surplus produce to a distributor from a nearby town:

> There is a wholesaler in Bornos, and he took the green beans off our hands at a really cheap price. What a surprise when we went to the shops in the *pueblo* and saw our beans, and the shopkeepers confirmed that the wholesaler had brought them to the *pueblo*. They were nearly triple the price we had sold them at! So we said: 'Enough. You'll never sell our beans again. You will never make money at our expense'. And obviously this is what led us to realise that we had to do direct sales. We would never sell again through a wholesaler because the only person who makes money is the wholesaler. The shopkeeper loses, the consumer loses, and above all the farmer loses.

Here we have a clear rejection of the open economy. What alternatives are available and how resilient are these alternatives? Supplying food to the *pueblo* gives La Verde their 'first and most important market'. However, their ability to sustain themselves through local sales is open to question. As cut-price supermarkets with a wide range of goods increasingly penetrate towns and villages, and local shops close down, they experience strong competition from 'convenience' stores. A possible alternative lies in connections to people who share an interest in and a desire to establish economic circuits outside mainstream food provision based around supermarkets and mass production. The vision of an oppositional alternative to the mainstream system takes us beyond the local, the natural terrain of the peasant economy, to links with actors at provincial, regional, national and international levels towards a description of wider networks and a broader politics.

Building Networks

Pueblos Blancos had been conceived initially as a way of bringing small organic farmers together to pool resources and develop a regional market that would lie alongside the export-oriented organic sector. This was successful for a time, in that the market expanded and Pueblos Blancos was able to offer a broader range of products to more customers than individual farmers or La Verde. However, the problems that led to the demise of Pueblos Blancos illustrate broader challenges faced by both small producers and alternative networks of food provision.

The contacts La Verde have long drawn upon include people in local and regional government, academics with an interest in farming systems, certification agencies, trade unionists, farmers groups, ecology networks, production–consumption collectives which grow food to eat, producer–consumer cooperatives that run retail outlets, private shops serving a middle-class clientele in well-to-do neighbourhoods, as well as wholesalers. The connections to these different people, groups and networks involve La Verde and Pueblos Blancos in processes that often contradict one another and lead to tensions.

At one extreme is the affiliation of La Verde to networks built on a social and environmental politics in opposition to capitalism. These networks are best exemplified today by radical groups of urban activists who have accessed peripheral land to produce and consume their food (López Garcia & López López 2003). The best known of these is the Madrid group of 'guerrilla gardeners', Bajo el Asfalto está la Huerta! (BAH!), but similar autonomous production–consumption groups based around horizontal decision-making also operate in Seville, Granada and Cordoba. Members pay a monthly quota to generate a fund, but also commit to work land they borrow, rent or occupy. The groups are popular – they have waiting lists – and they most closely represent the political ideal of collapsing production into consumption. On the other hand, the quantity and range of food they grow means that production for consumption must be supplemented by shopping. Hence it is necessary to distinguish between production for consumption and producer–consumer links, with the latter marking out different economic roles (producer and consumer), and so forming the thin end of a wedge into open markets. Discourse around local food sovereignty for the *pueblo* often loses sight of this distinction, and the same happens when applied to networks of people and groups.

Focusing on producer–consumer relations means returning to the market. Through links to farmers and cooperatives, La Verde and then Pueblos Blancos accessed products from other parts of Andalusia, Spain and Europe that they consumed themselves, or sold on through their distribution network (avocadoes and bananas from the Canary Islands and the Costa Tropical east of Malaga, milk from the Basque country in northern Spain, pasta from Italy). Qualities and practices associated with these products, such as mixed farming systems and the avoidance of waged labour and agro-industrial products, set parameters that bigger producers struggle to satisfy. In any case, these markets are not presently large enough to be of interest to specialists with economies of scale. To that extent, political ambitions are realised and reproduced through the network. However, the criteria and goals are complicated and often compromised by a second set of processes that coalesce around economic expediency. This occurs firstly because of the requirement to make a livelihood, and second because there are different kinds of markets and customers who are more or less committed to the political agenda. The way political and economic objectives play off one another is perhaps the most interesting thing about the networks and markets in Spain and elsewhere, and it raises the complex analytical issues we discuss in Chapters 8 and 9.

The problem of building and maintaining networks begins with internal relations between farmers within Pueblos Blancos itself. From the start, the aims and agendas of the politically motivated cooperatives were at odds with the ambitions of more individualist family farmers in Pueblos Blancos. La Verde ceded Pueblos Blancos its customer base when the latter group was formed, but the family farmers were reluctant to do this and tended to retain their own private customers. In turn, this led to shortages in products and exacerbated supply problems for Pueblos Blancos. For this reason the more political members accused the small farmers of being 'uncooperative' (*poco cooperativista*). Making an independent farmer president of Pueblos Blancos was a ploy to generate solidarity. But long running disputes over policy and practices led to recriminations that contributed to the demise of the group.

This outcome is part of a more general problem, that of the tension between economic agency and political vision. It is an issue that the more politically minded are keenly aware of, and is well illustrated by the case of La Verde's misadventures in export markets. The story began after a representative of a large French organic distributor appeared on their farm looking for produce to send north. Although this was successful on an economic level, with good returns and positive responses to their products from Parisian retailers, it meant they began to lose sight of their local markets, had to produce in larger quantities, and were subjected to strictures placed upon them by the distributor that threatened their autonomy. As Enrique says, 'they want to control you, so you become like their subsidiary'. The cooperative concluded export markets require unacceptable compromises. They, and Pueblos Blancos, insisted that the first priority was supplying local and regional markets, though they sometimes answered requests to sell surpluses to national distributors for sale within Spain. They have, however, objected to the global food distribution system and refused to sell for export. A member of La Verde explains:

> Food has to travel, I know. Globalisation means a complicated world, but what is not normal is that you can grow an apple here in the *huerta*, and they have to bring apples from Argentina. That is the strange, globalised world we live in, but for us it is not development and it is not sustainable.

While the values sustaining local markets are clearly defined and little contested – freshness, direct sales, the vestiges of a peasant-style economy, an ability to compete on price in a niche market – there is a range of ideas and commitments in the broader network that often complicates relationships. This was apparent in producers' attitudes to retailers compared to consumers. While farmers tended to identify with the latter and go to great lengths to indulge customers in the box scheme, or say things like 'the consumer is king', their attitudes towards retailers tended to be more ambivalent and judgemental. For example, preferential terms, prices and choice products were reserved for cooperative retailers and others with shared political visions, or with whom there was a long-term relationship. On the other hand, there was little sympathy for private shops struggling to turn a profit. I several times heard farmers insist that supplies should be cut to shops who failed to pay invoices on time. This is indicative of the price squeeze put on farmers, and the problem of liquidity within Pueblos Blancos, a second factor contributing to its demise.

Interviews at retail outlets supplied by Pueblos Blancos documented the ideas and commitments of retailers and their generally middle-class and well-educated customers. Here we meet politically motivated retailers and customers who are more or less radical in their demands and expectations. They adhere to an 'ideology of consumption', outlined in Chapter 3. This market is more difficult for big suppliers of organics to penetrate. But there are other retailers who sell mostly to people concerned about healthy bodies for themselves and their children. The latter customers want to engage in distinctive and ethical consumption, but may be more easily satisfied by generic labels and the cheaper prices of more mainstream products. The different expectations and commitments make the future trajectory of organics hard to predict, but the problems encountered by Pueblos Blancos and the financial difficulties faced by specialist organic retailers, both exacerbated by the recession, give pause for thought and further research.

For the moment the regional market for organic food favours smaller farmers. Large-scale producers, for their part, find more lucrative and bulk markets in northern Europe, and have not as yet paid much attention to the small, specialist, national sector. Most retailers reported a struggle in sourcing products and needed a range of suppliers. They expressed a preference for locally produced foods, but had to turn to national and even foreign suppliers. When it comes to fruit and vegetables, for example, they usually stocked products from Pueblos Blancos and other local growers, but also made recourse to Gumendi, a producer and supplier from the Basque country. Scarce, but sought-after products means premiums in the cities remain high.

Large national supermarkets stock few or no organic goods, and also charge high premiums when they do. So the picture for small producers growing an array of fruit and vegetables is of a niche urban market in which there is little competition and relatively high prices. In that respect it is more similar to the UK in the 1970s and 1980s, before big supermarkets stocked competitively priced organic produce. How long this will continue and the market remain viable for smaller growers is open to question, especially in the current economic climate, with recession and unemployment hitting Spain hard.

Urban consumers can purchase their organic fruit and vegetables (and other 'ethical' products) from producer–consumer cooperatives, from private shops or through box schemes. The first of these, the producer–consumer cooperatives, have both farmers and shoppers as members. Based in urban centres, they include La Ortiga in Seville, El Encinar in Granada and Almocafre in Cordoba. They have their own umbrella organisation, the Andalusian Federation of Consumers and Producers of Organic and Artisan Products (FACPE). Producer–consumer cooperatives have a small and politically committed group at the centre, with anti-capitalist and anti-corporate values. They run the retail outlets and decide policy. At the periphery is a larger general membership who may have political sympathies with the cooperative model, but are thought by the more radical core group to treat it 'like a shop'. If this suggests a lack of commitment, it also implies that customers will go elsewhere to buy food if they can get organic produce cheaper from supermarkets, or their incomes are squeezed by recession.

One motivation for being a member of these organisations is that prices are cheaper than in privately owned boutiques because cooperatives are eligible for subsidies, they are 'not for profit', and they get preferential prices from sympathetic producers. Privately run shops are also often run by committed individuals, but prices are higher and customers are generally less political. In the shops there tends to be less emphasis on social and ethical concerns, and more on individual and family health, issues that fall more readily within the neoliberal paradigm of individual consumer choice. Finally, there is the box scheme initiated by La Verde in the early days but later handed over to Pueblos Blancos, whose customer base includes ecologists, teachers and friends. Box schemes draw on political, ethical and financial motivations; they take advantage of direct relations between grower and consumer and the excising of intermediaries into an urban setting.

The array of outlets and consumers in the network undermines the idea of organics as a generic category and presupposes a range of relationships. Those relationships are personal, but were brokered by the distribution team at Pueblos Blancos (the manager, accountant, warehouse staff, and delivery team) who overlapped with and largely reproduced the political vision of La Verde. Dealings with different retail outlets were tempered by a number of factors. These included the extent to which there was a shared politics, the history of a relationship with La Verde and latterly Pueblos Blancos, and the volume and reliability of sales.

These preferences manifested themselves in the volumes and consistency of supply, and the different qualities of products that reached the consumers, whether through shops, producer–consumer cooperatives or box schemes. Shops tended to be a low priority, and shopkeepers often complained in interviews that they could not secure the products they wanted from Pueblos Blancos, and so had to source through distributors from other parts of Andalusia or national suppliers. The most important of the producer–consumer cooperatives to Pueblos Blancos was La Ortiga in Seville, an organisation with several hundred members. Key figures from La Verde were instrumental in its creation, so it is not surprising that La Ortiga, with its collective approach to decision-making, radical agenda and pursuit of direct relationships between consumer and producer, had fixed and preferential terms of trade with Pueblos Blancos compared to private shops.

The third group of customers are individual clients or consumer groups who participate in the box scheme as a form of direct selling. The most historically established of these customers are groups of ecologists who have had long-term relations with La Verde. During the time of Pueblos Blancos, the box scheme expanded to include a social programme organised and subsidised by the Andalusian regional government to supply schools, hospitals and old people's homes with fresh organic fruit and vegetables. Involvement in this programme was a double-edged sword. It allowed Pueblos Blancos to link up with like-minded producers in other parts of Andalusia, and so signified an extension of their network. Through this they developed a programme to share products, and so extend the range of foods they offered, as well as making moves to form a platform for lobbying and political action. On the other hand, there was limited success and some contestation in formalising the network. There was also wariness within the Pueblos Blancos' administration about becoming too dependent on official avenues for sales, as this diverts supplies away from other retailers. Further, their position within the social programme was seen as vulnerable to competitive tender, since large-scale commercial producers would probably win out.

While the network approach to distribution and sales appears likely to break open the closed system of the *pueblo* food economy, this is not necessarily the case. Rather, there was a conscious struggle to retain control of sales and make a virtue of the conditions of production. This was particularly so in the case of the box scheme, where the products were delivered direct to the end consumer and the producer–consumer cooperatives, who impose standards that exceed those stipulated by organic certification agencies. It is not, then, the distribution network itself that breaks open the closed system. Rather, involvement in organics as a commercial venture, opportunities in terms of price, the threat of competition that this brings, and the specialist and somewhat precarious nature of the market, haunts the network and potentially threatens smaller producers' livelihoods. For this reason, they always retain half an eye on their local customers in the *pueblos* as a kind of 'bottom line'.

Conclusion

This chapter has explored two strands in relation to local and organic food. The first, and most prominent, is the kinds of values that motivate and sustain people as they try to maintain and develop alternatives to mainstream food provision. The second, which is the conditions, limits and possibilities set by a more open economy, has appeared at certain points but largely remains the elephant in the room. In the grey area between these two lie tensions, which coalesce around how people negotiate and try to reconcile the money value of the open economy and their social and political convictions to pursue some degree of closure.

It is within this frame that the failed experiment of Pueblos Blancos must be examined. In part, this was a failure to compete in the open economy, especially in the current recession; shops failed to sell to consumers and pay invoices, and debts accrued on infrastructure and salaries. The role of subsidies in the story remains obscure. What is clear is that some kind of bottom line was set by the open economy. Alongside this was a failure of the network as it was conceived; relations between independent farmers and the politicised cooperatives were fraught and difficult to manage.

There is also the question of the wider networks and the potential or desirability of expansion. On a long drive back from one meeting I was party to a long and sometimes vociferous discussion between key members of Pueblos Blancos on the merits and pitfalls of expanding the network, as against consolidation of the local base. This debate was never resolved, but the default outcome was the latter course of action: a retrenchment back towards the local and a reinforcing of networks and contacts closer to home.

So, it is not a case of 'networks not working' (Edelman 2005); it is more a question of shifting alliances, moving sideways, and developing new initiatives. On my most recent visit to La Verde there was a new energy and determination to pay off the debts accrued from Pueblos Blancos. Extra workers had returned from the recession-hit tourist and construction industries on the coast. A new chicken shed had appeared and a bread-baking business was being developed, signs of determination and renewed vigour. The *pueblo* and local relations around food production retain their role as a site of resilience in hard times.

What, though, of the market and the open economy? The role played by certification in opening up organics to mainstream practice, capitalist assimilation and 'conventionalisation' is by now well documented (Guthman 2004a, 2004b). The principle argument is that big interests force down standards and outcompete smaller, more radical pioneers, who are then squeezed out of the market. The comment of one member of La Verde shows that the cooperative is well aware of the process: 'I am not in organics for a certification system which operates under shit rules modified to support the productionist sector. They accept dubious practices because of that, or they certify products that come from thousands of kilometres away'.

The challenge for smaller organic farmers and activists is to counter the conventionalising tendency by reaffirming the values that sustain the alternative: the social relation between producer and consumer and specific conditions of production that go beyond certification requirements. Manoeuvring between and beyond the bureaucratic strictures laid down by the European Union and implemented by regionally based certification agencies means ensuring customers see beyond the label to know where, how and by whom food is produced. This is understood by Enrique: 'we have never used La Verde like a label. We use the name and our prestige as a human group to sell'. For Manolo, it means 'returning to the minority', for example by breaking away from mainstream, third-party certification, and establishing forms of self-certification in which farmers monitor each others' practices. Experiments in such 'participatory guarantee systems' are already underway in Andalusia and elsewhere (Cuéllar Padilla 2010). As mainstream organics bends the rules to maximise profit, so alternative practitioners seek new grounds to differentiate themselves.

The second area that threatens the closed system is the market itself and the use of money as an impersonal exchange mechanism between people separated out into distinct categories of consumer and producer. When someone comes to the farm to buy, then the exchange is immediate and direct. When they send boxes to ecology groups, the exchange is less immediate but remains direct. Through shops, the exchanges are not immediate, not direct, and so are viewed as problematic. Money is the technology that allows the distancing of relations of exchange, but it is not the cause. For advocates of alternative food, problems come more from the separation of people into distinct categories of 'producers' and 'consumers', the distancing between them that money allows, and the capacity for entrepreneurs to profit from that.

In Andalusia an important model against which this is set is the idea of the closed, autonomous economy of the *pueblo*. This assertion is backed up by the emphasis La Verde puts on the *pueblos* as their first and most important market, and by Pueblos Blancos's preference for short, local circuits of exchange. Food provides a particularly clear window through which to see the strategies and values people pursue, the forces that draw them in to the open economy, and the problematic relations that emerge from that process.

Sources

The information in this chapter comes from fieldwork in Cadiz province over the last eight years. Data collection involved interviews and engagement with growers, retailers and consumers of organic food. It was made possible by many organisations and individuals, only a few of whom can be mentioned here. Fieldwork and subsequent analysis was generously supported by the Economic and Social Research Council. Martha Soler opened many doors, arranged interviews, and shared some of the fieldwork. David Gallard introduced me to the fieldwork site. The Universidad Rural Paolo Freire offered invaluable support. Members of La Verde cooperative were generous with their time, and in sharing their lives and work in the fields. The farmers and administrative staff at Pueblos Blancos, and especially Maria Carrascosa, were open and helpful beyond any reasonable expectation. The study would never have happened without the openness and generosity of all the people encountered during the research.

There is an extensive literature on anarchism in Andalusia (Brenan 1990; Corbin 1993; Foweraker 1989; Kaplan 1977; May 1997; Mintz 1982; for an overview, see Pratt 2003). Accounts of agrarian change, and particularly work regimes, *jornaleros* and their culture, can be found in Malefakis (1970) and Martinez-Alier (1971). Information on the organic sector and recent political and social movements that focus on food is largely in Spanish. The Department of Agriculture and Fisheries of Andalusia's regional government has published useful statistics on the organic sector (Junta de Andalucia n.d.). Del Campo Tejedor (2000) gives an overview of organic agriculture, and considers cultural aspects. In a recent edited collection, Guerrero Quintero and Soler Montiel (2010) examine rural transformations and emerging cultural and political initiatives around cultural heritage, agro-ecology and attempts to generate new agrarian models. Finally, work in an activist vein on social movements and food can be found in López Garcia and López López (2003), as well as in Autoria Colectiva (2006).

7

Sussex, England

Sara Avanzino

This chapter is an account of an attempt to localise networks of food production, distribution and consumption. It explores the relationships between producers and consumers as these emerge in various initiatives (two farmers' markets, farm shops and vegetable box schemes) in the town of Lewes in south-east England.

Interviews with farmers suggest they perceive the localisation of food systems primarily as a livelihood strategy, offering an alternative to mainstream food provision channels, in a climate of economic hardship which has worsened over the last 20 years. On the consumption side, the meaning that food sold as 'local' acquires in food outlets needs to be contextualised in relation to the increasing discursive significance that ethical consumption acquires as a practice. Buying 'ethical' local produce cannot be thought of either as a purely political act, or as mere concern for personal well-being. Neither is it solely a form of conspicuous consumption.

The following analysis of pricing issues that emerge at farmers' markets reveals that while there seems to be an underlying calculating rationality motivating actors involved in local food networks, we should not see these issues as just economic concerns. Both producers and consumers are found to negotiate between the dictates of the money economy and other values. The tensions emerging from this process lie at the heart of the chapter.

Lewes

Lewes is a town of about 16,000 people. Its origins date back to Roman times, and the building of the town's castle is attributed to the Normans in the eleventh century. The long history of the town attracts groups of tourists in search of cultural heritage. Lewes is located within the boundaries of the idyllic South Downs National Park, which confers an image of authentic rurality on the environmental landscape (Godfrey 2002: 122–23; Howkins 2003: 198). Over the years, and increasingly since the 1990s, Lewes and the south-east has attracted not only tourists but also city dwellers (notably Londoners) who have decided to maintain their urban employment and related lifestyle, but who have also chosen to live in a more rural setting (Howkins 2003: 208). This trend has contributed to making Lewes the prosperous city it currently is. It has also increased pressure on the surrounding countryside, and on the agricultural and farming sectors based there. I will return later to what significance this aspect of the demographics of Lewes has on attempts to localise food provision.

Lewes's history as a market town is another part of its cultural heritage. Early documents mention a daily market in 1089, whereby 'William II de Warenne gave to the monks of St Pancras, Lewes, the right of pre-emption, after the lord's needs had been satisfied, in this daily market "of flesh and fish and all other things which they wish and require to buy for their own needs or those of their guests"' (BHO 2012). A weekly market held every Saturday is documented from 1440. In addition to general markets, there were also specific ones for the trade of corn, livestock and wool, which developed from the late eighteenth century. The locations of these various daily and weekly markets changed through time, but a currently existing building, Market Tower in Market Street, remains a symbol of earlier events.

In the early 1920s, the Town Hall adjacent to the tower was leased by the East Sussex Federation of Women's Institutes (WI) to create a space where small producers of garden produce, poultry, eggs, jams and pies could sell their surplus directly to town dwellers or exchange it among themselves. That market is recognised as the very first WI market in the country. This location is now occupied by the weekly farmer's market, held on Fridays. The continuity between the WI market and the current one has clearly been interrupted by decades of structural changes in the farming and agricultural sectors, as well as in systems of food provision, yet it is a historical detail that the organisers of the current farmers' market give prominence to in their promotional campaigns and in their reinvention of the meaning of local food.

This weekly farmers' market was started in July 2010 by the Transition Town food group of Lewes as a project reliant on volunteers' free time. Now it is run as a community-interest company, a business independent from Transition Town Lewes. Despite being a separate body, most organisers nevertheless come out of the ranks of the Transition Town group, and share similar views about the need to localise food provision against the background of approaching oil scarcity and the unsustainable levels of carbon emissions caused by high levels of imported food. 'Creating an alternative food outlet has always been one of our aims', they said, 'we cannot steal people from Tesco [...] but want people to do all their food shopping here'.

Stallholders are selected on the basis of locality (a maximum of 30 miles) and their growing and rearing methods (certified organic or traditional), but overall they are 'expected to be looking for alternatives to chemical fertilisers and pesticides' (LFM 2013a). They include five vegetables growers (all certified organic), two apple growers (one conventional, one certified biodynamic), and three meat sellers raising their animals according to traditional methods. There are also secondary food producers: a baker, two cheese-makers and others selling homemade pies and preserves. Sometimes they are joined by others selling chutney, vegan cup cakes, ice-creams and household plants.

The other farmers' market has been running for longer but on a monthly basis. The first pilot took place in October 1998, and it is said to have been one of the first in the UK, after Bath. It is managed by another local organisation, the Common Cause Cooperative, established in 1991. The aims of Common Cause are broader than food distribution, extending to the establishment of housing cooperatives, the organisation of agricultural training centres, composting schemes and food growing initiatives. This farmers' market takes place on the first Saturday of the month, in Cliffe pedestrian precinct in the heart of town. There is space for more stallholders than the weekly farmer's market, and among them – the organisers enthusiastically emphasise – there are many producers who have won local, regional and national awards for their produce, raising the overall quality profile of the market.

Despite some disagreements, the management committees of the two markets seem to share the same ambitions and strategies. Firstly, both groups represent consumers' interests. They work towards bettering people's consumption experiences by pushing for more affordable 'local, fresh, healthy, seasonal food', as well as by providing them a place to meet those producing the food they eat and find out what happens out on the farm. Secondly, the organisers claim to work as representatives of local, small farmers' interests. The markets are designed to provide a space for local producers to network with each other, and crucially to 'test the market', beyond supermarkets. Farmers' markets are stepping-stones to launch new products, attract new customers and for overall business promotion. An analysis of how these various interests coexist and are negotiated in the marketplace is at the heart of this chapter.

Who are the producers taking part in these initiatives? The group is heterogeneous and not easy to pinpoint. I interviewed people working on eleven farms and for one food processing business, all located within a 19 mile radius of Lewes. They include three farms devoted to market gardening and four to livestock, three apple orchards and one dairy unit. Half of them work under some form of certification (mainly organic, through the Soil Association, the largest organic certifying body in the UK, the exception being one biodynamic apple orchard). Some, notably livestock farmers, refuse to certify, complaining about the burden of bureaucratic requirements, while maintaining what they define as a more 'traditional' approach to farming. Among other things, this also means rearing older breeds of cattle (Howkins 2003: 198). The people interviewed were on average in their mid 50s; as elsewhere in Britain, few young people are coming into the profession. More than half of these food producers have a family background in farming, going back one or more generation, with the notable exception of the market gardeners. In that group, two out of three are first-generation farmers who decided to get into organic farming out of ecological concerns with the damage which conventional farming causes.

Many of the farms employ some wage labour, partly because even the family farms have trouble retaining their own children in this low-income profession. In addition, an important factor that characterises organic farming and differentiates it from conventional approaches is the opportunity to attract volunteer (free) labour through organisations such as World Wide Opportunities on Organic Farms (WWOOF), which was founded in 1971 to bridge the gap between increasing urban demand for countryside-related experiences and the lack of rural manual labour. One organic market gardening outfit and the biodynamic apple orchard also benefit from the work offered by students of a biodynamic agricultural college, located 20 miles from Lewes and populated by an international community (including Dutch, Italian, Spanish, Indonesian, German and South African students).

Living 'the Local': Actors and Issues

The aforementioned two local markets reflect a wider interest in food and its origins. At a national level, there are several organisations, associations, foundations and government bodies in the UK that work in the discursive field linking community development and food production and distribution. An example of research carried out on a national level is that of the Campaign to Protect Rural England (CPRE), which aims at making 'local food webs more visible and better understood – to put them literally on the map – and make clear their ability collectively to make a difference' (MLFW 2008). The final report draws on findings gathered by community research groups across the country, and opens with a problematic attempt to define what local food is (CPRE 2012: 10–11). What do we need to know about a place to understand its local food economy? Let us go back to our case study.

Some of the farmers I interviewed remember an earlier system of food supply. In the 1960s, Brighton, a conurbation 9 miles from Lewes, was fed by a belt of market gardeners located in the surrounding countryside, organised through a network of wholesale markets which supplied the city's greengrocers. Such marketplaces still exist, I was told by a conventional apple grower who trades there, though on a much smaller scale. Similar stories are recalled in the livestock sector. Michael and his family have been working their farm for five generations, first as tenants at what was a dairy unit, until Michael's great-grandfather bought the land after the First World War and turned it into a livestock farm, rearing chickens and lambs. During one of our conversations, he and his mother recalled how up until the 1970s they used to sell mainly to independent shops in the area. There were more independent butchers' shops back then, and the meat traded in those outlets constituted the bulk of livestock farmers' trade. From the late 1970s onwards, many of those independent shops closed down as supermarkets gradually gained a larger share of the national food market. Similar stories are recounted in the fruit sector.

Although the rise of supermarkets did mark a crucial turn in the way food is distributed in the UK, it was not a one-off event. We need to see this as a series of economic and cultural changes that developed over a long period. Supermarkets have not always had the same buying power and the same centralised structure. Michael and his mother could recall a time when they used to deal with the Cooperative chain, through their independent butcher departments, which sourced more locally and paid the same price as local butchers. Likewise, a couple running an organic apple and pear orchard commented on how supermarkets created and supplied a demand for highly aesthetic produce. This affected, and still affects, producers by forcing them to comply with very strict requirements. They argue, 'the problem is not with the specifications themselves, but with the fact that they don't pay higher prices for their strictness [...] whereas in the past they did'.

The food distribution networks in place before the establishment of supermarkets has withered away since the 1980s; it has not disappeared, but it has considerably diminished in size. There are still wholesalers, but not enough of them, as some growers lament. What this means is that since the mid 1990s those farmers and growers who refuse to deal with supermarkets have had to find other markets, or create them where they did not exist. Farmers' markets have provided one possibility; vegetable box schemes and farm shops are other examples. Importantly, although they sometimes benefit from an image as traditional remnants of the past, of 'old-time' country markets, they need to be recognised as quite recent innovations, crucially so because they represent a choice for both producers and consumers, and coexist with other channels of food distribution.

Efficient or Inefficient? It's Not Just a Matter of Scale

In many of the conversations I had with farmers, they raised the issue of efficiency. To delve into the significance it has in small-scale farming, we need to consider the various factors that influence not only the commercial viability of those kinds of farming operations, but also the overall aims of the people involved. What is it that producers want to achieve in running the farm, apart from maintaining their financial viability? Efficiency has to be understood in these different but coexisting dimensions. Let us now explore how this negotiation takes shape in practice. Adrian explains:

> I think it's scale, if anything. Even though we've got a niche market being organic, it's still scale that is getting us the major difficulties. We grow a wide range of products; we haven't got the economy of scale to be able to specialise. This makes our margins quite tight. It amazes me how we are able to compete, if you think about how small we are compared to the scale on which most vegetables are produced.

Adrian is a first-generation organic market gardener growing produce on 15 acres of land, and retailing directly with customers at farmers' markets and through a vegetable box scheme, which he has been running almost since beginning in 1997. His preoccupations are shared by many other small farmers and reflect the difficulties they face in securing a market for the quantity of fruit, vegetables and meat they produce. Being efficient, in these terms, expresses a relation between the ability to produce and the ability to sell. It is an economic calculation which needs to be made, as these farms do not operate as closed systems and self-sufficient units: they rely on the market in order to procure some of the inputs for farming (seeds, fodder), as well as in order to obtain capital to reinvest in the farm, pay wages and supply farmers' personal consumption needs.

The choice these farmers make to engage in the localisation of the food economy in south-east England still means that they are deeply embedded in a market economy, at different levels. Farmers' markets, farm shops and box schemes are all commercial operations that do not exist in a vacuum, but rather represent a choice, among other food provision channels, not only for producers, but also for those customers who want to source their food according to criteria of locality, sustainability, freshness and quality. Emblematically, walking along the main street that hosts the farmers' market the first Saturday of every month, one crosses, behind the stalls of local producers, other food retail shops and supermarkets, all trying to appeal to customers and competing for their custom. Issues of efficiency spring from the need to maintain the farm as a financially viable operation amidst competition from the integration of global food markets, of which supermarkets represent a powerful arm. Adrian observed that, 'if the price of oil goes up, we become more efficient, because our products will cost less than the ones more heavily based on the oil economy ... without doing anything different'.

It is not only world oil prices and international labour markets that influence English food prices, but also the scale at which that food is produced. As we learned in Chapter 2, the majority of the farms that provide food for northern markets operate as large-scale units that depend on heavy levels of mechanisation and monoculture. At the same time, the buying power of supermarkets and consequent centralisation has grown rapidly. The consequence of such scaling up, in production as well as in distribution, has been paralleled by a loss of producers' ability to set prices, as they become dependent on a market able to regularly buy in bulk. As Carol, a livestock farmer rearing traditional breeds of cattle, observed:

> The problem with selling wholesale is that the more efficient you get, the more supermarkets want to push your prices down. That means you have to get more efficient to compensate for the diminished returns, and it ends up in a constant down-spiralling of prices.

The crucial point is that down-spiralling affects all producers. Once prices applied to large units of production set an international standard, their efficiency (as calculated in terms of production costs) will inevitably impact on smaller units. Farmers' markets are very much part of these structural dynamics. There is no perfectly efficient balance for smaller-scale farmers to achieve, a set blend of land, capital, labour, machinery and customer base, as the combination of these variables is not just a technical matter. As many of them have commented, at some point in their life they found themselves at a crossroads, pushed by global restructuring; either they had to grow in size, specialising and mechanising, or remain small, more labour intensive, and find ways to add value to their products. Or find a compromise between the two.

The story of that choice is the story of the struggles these farmers are going through. No one farmer had exactly the same answer to give about what is the most viable strategy, for those choices are generated out of personal life stories, passions and values. All of their stories lead us back, through different routes, to the notion of efficiency, and force us to rethink its meaning. The question to be asked then is not whether small-scale farms are efficient or not, but rather what is the notion of efficiency they are working with? Is it at producing vegetables, apples, eggs and meat at the lowest unit cost possible, or at doing something else, or more?

How Do They Grow? The Impact of Scale on Methods

The thorny issue of efficiency arose during a conversation with Collette, a first-generation organic market gardener:

> people like us are considered inefficient, but that's when you don't include the environment and animal welfare. There's something that gives you an immense job satisfaction; having your own piece of planet you are a guardian of and you look after. You can get to know it really well, it's good husbandry.… [In the past] they used to see land and soil and the environment as something a bit more alive and tricky than a mere production unit, as economists would see it. That for us is the core thing.

The scale of production, in most of these farmers' views, is more than a mere extension of land or a number of outputs. Matthew, who turned his grandfather's conventional apple orchard over to organic, understood that 'converting to organics doesn't mean to only convert your land, but also to change the way you think about growing and selling'. Only half of the farmers I spoke to grow organically, but all share a sense of respect for the land and animals that underlies their operations. This informs their understanding of efficiency and problematises its classical economic definition. Market gardeners and apple growers agree that feeding the soil and not the plants through crop rotation, the use of green manure and compost, and enhancing biodiversity through companion planting, is the most efficient way to build up soil fertility for the long term and to grow stronger plants. Such techniques deliver lower yields but less susceptibility to diseases. 'Conventional growing depletes the soil of everything; it becomes a matter of water and fertilisers'.

Similarly, in the livestock sector, rearing traditional breeds of cattle, sheep and pigs means rearing animals that have not been selected for their 'productive capacities', that is milking and fattening; it means not forcing animals' growing cycles, not feeding them with high-protein feed, and not killing their instincts. As Clive said, 'the natural way of doing these things is also the easiest. Traditional English breeds tend to need less human intervention and are less susceptible to illness. In a good year, sows require only about 10 to 20 per cent assistance [during birth]'.

'Being in control' is a concern farmers mentioned often. In some cases, it refers to inputs. As Adrian said, growing vegetables organically implies, among other things, refusing to buy into the fertiliser industry, which 'would dictate how I grow' by enforcing certain methods and frequencies of use. The notion of control is also mentioned in terms of the ways in which farmers choose to sell their products. Retailing to customers without intermediaries and remaining in direct contact with them enables producers to do two things which increase their control. Firstly, they can persuade customers to be more flexible in their choices by getting them to understand farming dynamics better. Secondly, by getting direct feedback from customers, farmers can sense their preferences better, and they are able to choose how to accommodate them. This means not becoming dependent on anyone else's marketing strategy.

There is another context in which the term 'control' occurs, with rather different connotations. It is best illustrated with their own words. Matthew, an organic apple grower, explains:

One of the key words in conventional fruit growing is 'control': controlling pests, controlling disease, etc. [...], [w]hereas, in the organic world, you really have to take the breaks off the control factor and have a leap of faith in nature. I found that natural predators will control the pests, to an extent. You won't get total control, but you have to accept that there is going to be a certain amount of damage, because that's the nature of the beast. The pests–predator relationship is such that the pest has to be present to a certain extent for the predator to be present; nature affects the control. My principle is that you don't do anything to upset that, that's your primary control measure.

This does not mean advocating a totally passive relationship with ecology. Rather, it points to a different conception of work, one closer to that of craftsmanship and artisanal independence than industrial, mechanical production. Adrian describes his job as a trial-and-error process, and it is this creativity that motivates him to experiment with new techniques or plants, invent new solutions, and to keep on farming. Farmers and growers are keen to describe their job as a craft, requiring inventiveness and responsiveness to unpredictable conditions, whether climatic, agricultural or, not least, economic. Depending on fertilisers, Adrian argues, 'is like painting by colouring boxes'. Driving back from a day's work at the farm and asking about what makes good farming practice, I was told that 'the difference between a good farmer and a bad farmer is two days'; the speed and the quality of the responsiveness to problematic conditions is crucial for obtaining a good harvest. In this light, independence seems almost an indisputable requirement for good farming practice, and it sheds some light onto why it is valued so much. As Toos, a Dutch organic market gardener who obtained the fourth ever Soil Association certification in England in 1978, put it: 'To be my own boss is the most important thing: it's a challenge, a very demanding job, but I feel so free. I am in charge. I love it.'

These farmers know that there is a certain scale of production at which all this changes, a level at which methods of producing must alter, and with it the non-physical, non-financial returns farmers obtain. That boundary is quite blurred and is not the same for everyone. Colette describes it as 'when it stops being a craft', Toos as 'when I can't control all the work myself', and Sharon as a point at which you stop receiving adequate returns. Yet, every one of them is aware of it and is careful about not crossing that line. Matthew eloquently expresses the tensions: 'you have to resist the temptation to get bigger. Some people can't resist it and grow and grow. But they don't do themselves and anybody else any good'.

Whatever the ethic, all farmers have to face the necessity of securing a market for their products. Producing is only half of the story. In conversations with Adrian, for example, while discussing the efficiency of his farm, the meaning of the concept kept shifting – from soil fertility to economy of scale, from productivity of the land to tradability of the products – demonstrating the tensions that emerge in the process of negotiating personal values with financial needs. 'It is because we are retailing that we are surviving on such a small scale'. So what does it mean to retail for these small-scale farmers, and who do they sell to?

Selling Products: Adding Value and Marketing Options

Direct retailing increases returns to the farm, and is one of many strategies that add value to farm products. Understanding the process of adding value and how it develops in particular economic and social settings is crucial to grasping the trajectory of this farming sector. The three main issues we need to look at are the motivations, the methods, and the consequences of such value-adding strategies. With regards to the first, the choice to add value stems from a need for small-scale farms to find ways to survive the low prices they would otherwise receive from large distribution chains. Larger farms work on economies of scale, getting diminished returns for each head of cattle or kilogram of vegetables but aiming at getting sufficient returns by trading large quantities. Small-scale farms could not survive on those prices for the quantities they produce, nor could they produce the amounts supermarkets would demand, even if they chose to deal with them.

With regards to what methods are adopted to add value, we need to differentiate between two major trends – though they are not mutually exclusive. As we have seen, value can be added by acting upon farm-related activities, by taking charge of the retail process, or by transforming the raw materials and increasing the value of the final products, for example by making preserves or fruit juices. In addition, it can be done by diversifying farm resources out of food production, for example renting out fields as pony paddocks, turning glasshouses into caravan storage facilities and barns into houses.

These two trends have important consequences. With regards to diversifying farm resources, it is worth noting that the UK government is actively promoting this as a positive developmental step for the farming sector. What this will do, though, is externalise the structural problems of the farming and agricultural sectors into other areas of society, because agricultural work itself is not rewarded and stimulated. It hence exacerbates the gap between urban and rural employment and ultimately calls into question the meaning of rural livelihoods themselves.

The other value-adding strategy has, instead, a different kind of consequence. Retailing directly to customers is, in fact, not just a matter of cutting out trade intermediaries. This in itself would not guarantee the success of the initiative. It also requires customer engagement. Farm retailing is heavily affected by the process through which the social significance of 'buying direct' becomes inscribed into a monetary value, a premium price that certain people are prepared to pay. As Colette declared, 'we cannot just sell our vegetables, we have to sell an entire belief system'. This includes the importance of defending sustainable farming practices, supporting local farmers against the pressures of supermarkets, and the need for small farmers to obtain a decent livelihood. The construction of that social significance is part of a multi-layered process of differentiation. On the one hand, 'locality' as a quality defining both producers and products remains in competition with other food values (such as affordability, taste, organic provenance) in the marketplace for consumer choice. On the other hand, consumers themselves recognise their choice to subscribe to a box scheme as informed partly by class dynamics and some forms of cultural capital, which distinguishes people according to the priorities they adopt in their food shopping. However, there is a twist. One of the consequences of turning lifestyle into a saleable product is the opening up of possibilities for larger farms and dairies to appropriate these value-adding strategies, selling under the 'local ticket', and acquiring higher returns despite not being small farms.

This happens to an extent at the farmers' markets in Lewes, but again we cannot operate a black and white distinction between those who sell directly to make ends meet and those who are merely looking for higher profits, for at least two reasons. Firstly, even medium-sized farms are struggling, although their struggle is more oriented towards securing a livelihood than living up to political principles. Secondly, even those farmers who seem more committed to the localisation of food economies have reservations about what they do. Or rather, they have reservations about the way in which the overarching discourses about food affects what they do. As one producer put it, he is forced to find ways to add value to his products, 'marketing them as extra special, so that you end up selling Christmas presents rather than food'. There are discursive constraints within which producers are forced to operate. Expectations of cheap food supplies mean that producers have to justify their higher prices, and this sometimes conflicts with their ethical principles.

Let us disentangle this process of adding value further by looking closely at the wider marketing strategies of producers selling at markets and farm shops. One of the principle observations relates to certification. Why do the majority of organic producers choose to certify with the Soil Association, the largest organic certifying body in the UK? Most of them explicitly recognise how the famous logo opens commercial doors that would remain closed if they did not adhere to an internationally established set of regulations. In this case, the bureaucratic burden of certification remains an inconvenient requirement, but it is repaid by considerable final added value. For some, certifying represents the enforcement of an actually useful set of practices. Yet for others, who refuse to certify, it represents a real financial obstacle, or even an ethical dilemma. They contest the practices and costs that certification imposes. According to Michael, a livestock farmer, it is more important to make sure the animals have a good life, have adequate space and live according to natural growth rhythms, than whether they eat organic feed or not. Similarly, Sharon, another livestock farmer, commented how she prefers her customers to buy from her farm shop based on personal trust and direct knowledge of her farm, rather than letting her practices be assessed by an external certification body. Across different agricultural sectors, certification remains a contested terrain. It undeniably serves, though, as a value-adding tool that turns certain farming practices into a saleable product, in distant as well as in local markets.

By exploring producers' marketing strategies, we have also learned about the strengths and limitations of farmers' markets as value-adding venues. None of the producers I met sell exclusively at farmers' markets. The reasons vary, and it is through this variation that we can get a clearer grasp of the diverging kinds of ethos that keep people producing food and selling locally. On the one hand, we have farmers with highly diversified production regimes (mainly market gardeners), who combine farmers' market sales with local box schemes. They lament that their sales at markets are not substantial enough to absorb their entire production costs, and sometimes do not even justify the two days of work they generate.

On the other hand, at farmers' markets we find a different kind of producer. These specialise in one or two products (apples, eggs, cheese) and rely on other regional or national markets. In addition to supplying local outlets such as restaurants, box schemes and farm shops, some sell through wholesalers, others also to supermarkets. Some of this group of specialist producers also express reservations about farmers' markets. Their products are increasingly available in local shops, so as producers they can reach their customers for potentially similar returns without the trouble of spending a whole day behind a stall, depending on the volume of sales to shops (where margins are tighter) compared to that on market stalls. Many keep on doing it because 'it's a good PR exercise'. The scale of production of this latter group is usually larger than that of those selling only locally. For them, local markets (including farmers' markets, farm shops and health-food shops) constitute only a small portion of their overall trade, but for some of these producers these outlets nevertheless provide the most substantial proportion of their overall income. As a producer growing apples conventionally on 40 acres of land told me, he gets 60 per cent of his income from the 20 per cent of his overall production which he sells locally. At farmers' markets he sells apples that supermarkets refuse, because they are slightly damaged or are too ripe and would otherwise go for juicing, and from which he would only get 8 pence a kilogram instead of the £1.20 he charges at farmers' markets. Experiences in cheese production are very similar. The returns that one producer I interviewed gets from supermarkets only cover the fixed costs of production (wages, bills, maintenance), and it is the revenue from local retailing that he regards as profit to reinvest in the business. By buying large quantities, supermarkets allow makers to maximise the efficiency of their machinery, hence increasing their economies of scale, which in turns makes them more competitive in the local market too. These two cases are representative of a number of producers selling at farmers' markets, and their strategies reveal the less visible reality behind discursive representations of 'noble farmers', committed only to feeding local people and not interested in profit.

The point to emphasise here is that farmers' markets, and the products sold at them, represent niche markets. Some of the producers have reservations about the direction that the movement for the localisation of food networks is taking, and the amount of profit being made out of it. Others are happy to exploit commercial opportunities. However, faced with fierce competition from the mainstream in terms of production costs and retail prices, all of them choose or, perhaps better, are forced to participate if they wish to survive. The opportunities offered by the niche market are an outcome of the particular kinds of customers that frequent farmers' markets, and it is to them I now turn my attention.

Customers

Matthew's wife Carol commented on their customers, saying, 'you are an independent shopper because you like independent shops'. Who are these 'independent shoppers', and what are the values they want to realise? They are interested in expressing their ethical choice through the market, but a price-based mechanism can reveal contradictions.

All farmers are aware of the general level of wealth in south-east England, and particularly in Lewes. When Clive stopped getting subsidies for his livestock farm, he started to charge his customers for door-to-door delivery, but 'they didn't seem to mind'. His customer base, he says, is different from that of the local butcher: 'my butcher in Burgess Hill buys in whatever is cheap because he's got that sort of market where people respond to prices. He looks at my meat with horror. He could not possibly sell meat like that'.

Prices for meat and eggs from traditionally reared animals, and for organic vegetables, are higher than conventional products for multiple reasons. The issue I want to examine here concerns farmers' politics in relation to higher food prices. How do the producers selling at farmers' markets combine their efforts to localise food with the fact that the majority of people are unable or unwilling to pay the prices? Is it seen as a contradiction? Such questions open up a series of debates.

The answer to the issue of affordability is not straightforward, as there are many factors determining people's food consumption choices apart from their financial means. In the same way, it is not only the search for income that motivates farmers, despite their need to generate a livelihood. The subtleties of the process of negotiating these elements are not easy to describe. What remains clear is that the wealthy customer base found at farmers' markets is seen as an advantage and not as a problem when it comes to spreading the model of a more localised food economy. Farmers might not necessarily try to generate high incomes, but they accept the fact that the money they receive comes from a restricted social group.

When discussing local food affordability, most of the farmers lament how the debate should not be confined to matters of price, but insist collective understandings of affordability are also heavily shaped by customers' priorities in choosing where to spend their disposable income. As Michael points out, the proportion of income spent on food has on average halved over the last 20 years, from 20 per cent to 10 per cent. This is the result of two processes. First, it is a reflection of the lowering of food prices, following the global integration of food markets and because of supermarkets squeezing suppliers' margins. As Colette argues, complaints about high prices need to be contextualised in a society which expects, and has normalised, unsustainably cheap food. Second, farmers point out that the drop in food expenditure is also a result of the fact that food, flavour and the skill of cooking are not much appreciated in English culture. Hence, the argument goes, people are more attracted to the cheap and convenient food supply provided by supermarkets than understanding the conditions under which the food they eat is produced. Those who decide to buy at farmers' markets and subscribe to vegetable box schemes are seen as a minority who have recognised and agreed to pay the 'real' price of food. The middle-class clientele are transformed into 'the community' supporting local, small-scale farmers.

If the debate about local food is not confined to matters of price, what other elements constitute it? The values consumers want to realise are heterogeneous, and sometimes contradictory. Adrian asserted, '450 customers have 450 reasons', but some recurring themes can be traced. Observing the changes occurring in their customer base, some of the farmers admit that first 'organic', then more recently 'local', have become labels, selling points and marketing strategies. Consumer choice, they lament, swings according to the shifting popularity of ethical themes. 'The campaigning [that] supermarkets did to promote their own "ethical line" influenced the way our business appealed too', one farmer commented. Some vegetable box customers, who said they were initially more interested in the organic quality of the food, now stress the importance of provenance. 'I think consumers are quite confused', another farmer remarked. Or as Colette says: 'The disadvantages from our point of view are that we are both growers and retailers. We have to do two jobs for the price of one – which is a lot of work [– and] while we had our heads down growing the stuff, we failed to notice the market's motives for buying organic' (BHFP 2012).

Another theme is revealed in a set of interviews on the wider shopping habits of those who subscribe to vegetables box schemes. In most of the comments made by vegetable-box-scheme customers on the quality of food in their boxes, priority is given to taste, freshness and higher nutritional value. The authenticity some of the customers describe is explained in terms of good flavour, of what food 'should taste like' in comparison to the blandness they get from supermarkets products. These nutritional qualities are often translated as expressions of care for their children's well-being. Further positive value is attached to the strictures imposed by the range of products in the box. Some admit that a seasonal diet is challenging, but this forces them to rethink their cooking habits, requires more creativity, and represents a positive challenge. Similarly, people comment on the surprise element of the box scheme, to the extent that 'every box feels like Christmas' – and this sense of surprise contributes to the overall significance of box schemes and to the decision to subscribe to one.

The content of the box does not usually satisfy a family's food requirements. Among the people we spoke to, much of their shopping is sourced from supermarkets such as Waitrose – which farmers see as one of their main competitors precisely because of its marketing strategy to sell under the 'ethical' label, but with a bigger range and at keenly competitive prices. The interviews show that, while people are shopping, the criteria for locality and seasonality is open and applied selectively to specific products for particular occasions. Ethical products most often represent only a small part of people's wider consumption choices, and they seem to be bought intermittently and in addition to other products, where the underlying basis of choice is price. In Chapter 3 we explored why food is at the centre of many people's ethical consumption. Overall, though, it seems that when people shop at farmers' markets or sign up for box schemes, they conceive of doing so as a choice, a means to 'treat oneself to some ethical products'. As one customer put it: 'I'm not consistent [in my criteria]; it depends on whether I can afford to be righteous or not'.

Behind this kind of ethical consumption there seems to be more concern about the quality of the food itself and of its affordability, rather than the social conditions of production and their implications for the wider farming sector. The lack of a clear definition of locality, and of the social and economic implications of more localised food networks, seems to indicate a missing shared social and political agenda. Ethical consumers are hence left to adopt their own individual and rather technical interpretations of locality. On the one hand, farmers' markets set their own radius: in Lewes, 30 miles for the weekly market, 40 miles for the monthly one (even if it is not clear how far down the chain the raw materials for secondary products have to comply). On the other hand, customers pledging to buy locally – at least for some food, some of the time – establish subjective boundaries too: Spanish strawberries are acceptable, Kenyan beans are not.

Pricing and Labour

Most customers report that they operate some form of price comparison and consider whether they get value for money, not only at supermarkets but also in subscribing to box schemes and buying at farmers' markets. Some of the organisers of the markets themselves encourage that rationale. A leaflet promoting the weekly farmers' market in Lewes invites customers to 'buy the best local produce at market prices'. Their reasoning is elaborated on in their webpage:

> we are running a market and inherent in this is the principle of customer choice. This means competition and change, if the market is to stay fresh. While the quality of the produce is an important driver, a real market also contains bargains. It is possible to have both; indeed it is essential to attracting customers. (LFM 2013a)

This points to the heart of the problem: price justice, defined as (customer) affordability compared against market rates. The influence of the neoliberal mantra of competition as a price leveller on local systems of food provision is reflected not only in customers' expectations, but also in farmers' practices. Many of them explain how their pricing involves negotiation between their production costs and supermarket prices. Regardless of whether their prices match supermarkets or not – most often they do not – the politics of value unfolding at farmers' markets are 'interior rather than exterior' (Kahn 1997: 76) to the forces shaping wider economic circuits. That is to say, in order to choose what price to charge for their products, it is not enough for farmers to calculate the real costs of production and reflect those in the final price. Instead, they are forced to consider how competitive they are in relation to larger food distribution outlets, and adjust their prices accordingly, even when they sell to the local 'community'. This is strikingly at odds with the supposed aim of farmers' markets and box schemes, which is to re-localise food economies to protect the livelihoods of small-scale farmers threatened by competition from supermarkets, whose commercial advantage is determined not only by their economies of scale, but also by exploitative relations of production they indirectly impose. Downward pressure on prices is then reproduced in urban consumer expectations about cheap food, which is often refracted through class (wanting to make locally produced food affordable for the less affluent strata). However, as Colette highlighted, those demands ignore the wider political economy that has normalised an unsustainably 'cheap' supply of food. 'If we charge more we are seen as elitist and anti-working class. But actually, no, we are the working class'.

Some producers insist this kind of price comparison makes it more difficult for consumers to understand the real conditions of small-scale farming. Yet they themselves, together with their supporters, propose alternative price comparisons, in relation to ethical lines in supermarkets. Organic vegetables bought at farmers' markets, they claim, are cheaper than organic vegetables bought at Waitrose, or even Tesco. Carol, an organic apple grower, finds this frustrating: 'people assume that Tesco's prices are necessarily lower than retailers', even when that's not the case'. Two comments help disentangle what is at stake in these economic calculations. Firstly, the fact that supermarket prices are in most cases used as the standard against which other prices are compared reveals the extent to which their discursive and commercial practices have elevated them to an authoritative position in terms of 'value for money'. Resisting this is no easy task. Secondly, it is revealing to note how much the terms of the overall debate on local food, as it unfolds in England, are informed by a calculating rationality and how often financial mechanisms are called upon to formulate arguments in favour or against the localisation of food economies, or even just to comment upon it.

In trying to understand the origin of these tensions, we need to remember that both farmers' markets held at Lewes are consumer-led initiatives. Two of the members of Common Cause Cooperative who contributed to setting up the monthly market in Lewes said they proposed that farmers be directly in charge of it, changing the operational structure from a social enterprise to a farmers' cooperative, thus giving farmers more independence in structuring it. The response they got was negative, as the farmers explained they were interested in selling at the market, but not in its organisation. On another occasion, farmers declined the opportunity to lead a farmers' networking initiative that Common Cause had kick-started, again preferring that someone else remain in charge of its coordination. Farmers themselves confirmed that, despite the existence of informal networks of mutual support, there is no formal cooperation. Protective attitudes to private resources and individual interests remain an obstacle to cooperation. Nor, historically, has there been any well-embedded political culture in UK farming. When discussing how the exponential increase in the price of land is precluding new generations from farming, and the lack of farmers political representation, Matthew commented, 'it's all going to happen by default rather than planning'; the localisation of food production will become a necessity and the system will adjust in turn. We should not generalise from one remark; nevertheless, the role of farmers' political organisations was almost absent from my conversations with them.

One of the consequences of the lack of political organisation is that all attempts to change the shape of the current food system are conceived in terms of relations between producers and consumers, and of their reconnection. That reconnection is then eventually expressed in terms of monetary transactions. Reconnection is a valuable step in reducing customer alienation by offering an embodied producer to sell them their food, happy to tell them how it was grown and where, but it does have less obvious negative consequences. By emphasising only one kind of social relation out of the plethora that informs these two (fictitiously defined) categories of people, farmers' markets discursively reinforce the divide it originally set out to bridge. In other words, consumers are encouraged to approach farmers' markets with consumers' expectations, and producers with producers' expectations. In the process, the fact that producers also contribute to the wider economy as consumers is obscured by the image of the friendly, generous grower. Of course this is a generalisation, but a few farmers express quite strongly their frustration at the difficulties they face in overcoming the producer–consumer divide, which they say is deeply rooted in people's psyche, even at farmers' markets. Poor consumer insight into farmers' needs, they add, is a problem for achieving a socially just local food economy. As Collette says, 'I don't blame the people. I blame the market economy'.

In practical terms, farmers report that this divide manifests itself in different ways. Firstly, in terms of the expectations that some of the organisers place upon the producers, such as to lower their prices. 'You'll find a wide range of competitively priced fresh produce so you can do your weekly shopping at the market, not the supermarket', we are told by the Friday market website (LFM 2013b). By insisting on lower prices, those organising farmers' markets risk distorting the origin of the problem. If, on the one hand, burdening a few small-scale producers with the task of competing with supermarkets helps to increase the accessibility of the markets, on the other hand it means denying the structural constraints those farmers are forced to operate in. The scale of local food trade needs to increase for it to stop being an 'ethical treat' and become the norm, but 'there just aren't enough producers' is the unanimous comment of farmers. Another example of the divide between producers and consumers is linked to consumer choice. Farmers lament how selling the idea of seasonal food sometimes becomes an obstacle because customers 'want consumer choice as if it is a human right, but most don't see the consequences of that choice [...] We end up living on what someone unemployed lives on, but most customers don't see that'.

What slips through the net in these attempts to develop a 'moral' local economy are the broader dynamics of the political economy of the agricultural sector in the area. For example, contrasting levels of wealth is a significant issue. It generates different visions of the countryside, which farmers argue is one of the main difficulties they face. The south-east has attracted a large population of Londoners, looking for a rural idyllic landscape to live in while at the same time maintaining their urban employment in the city and the lifestyles that their non-farming income allows. One of the major impacts of this trend is the increase in the price of farming land. When people from outside the agricultural sector pay much more for land than the income that can be extracted from it by farming, the next generation of farmers is excluded. It also discourages current farmers, who see the financial rewards of their activities belittled in comparison to the amount of income they could make by turning their land into pony paddocks or golf courses. This issue, though, remains invisible in the kind of 'politics of morality' and individual ethics that initiatives such as farmers' markets promote. That has the effect of depoliticising the so-called 'community' that supports local farmers, which, sometimes, is responsible in the first place for some of the structural problems farmers find themselves in.

Concluding Voices

Clive is a small-scale livestock farmer who stopped selling his beef and pork door-to-door a couple of years ago because of a mix of financial difficulties and demotivation, the latter brought about by the body-blow delivered by BSE and then foot-and-mouth disease. He repeatedly told me how selling to wholesalers and supermarkets meant being 'at their mercy [whereby] you cannot set your own prices', and how being small and retailing directly protects producers from the global manipulation of food prices by large companies. After having heard the voices of other producers, who struggle to secure even the equivalent of the minimum wage from selling at farmers' markets and through box schemes, we can no longer be so confident about that. The debate goes back to the question of how growers price their products. Toos, who comes from a Dutch farming family, once said of farming: 'it's a 24 hour job'. How can we reward the creative energies and passion farmers put into their labour? Surely, in a purely price-based mechanism something gets lost, and that something is what keeps them doing such a demanding job. That is why it is important that farmers' commitment and creative energy does not get lost in the practices adopted to localise food economies and that the movement moves beyond rhetoric.

Finally here is a market gardener, commenting on the nature of the recent revival of interest in food:

> The current craze about local food has seen a wealth of initiatives such as celebrity chefs, food partnerships, and a lot of leafleting. But actually nobody else has started growing vegetables around here, or converted to organics, at least not in the last twelve years I've been here. People get into things because they have to make a living out of it, but you make more money producing leaflets.

Sources

This chapter is the result of three months of fieldwork and of exchanges and conversations with many people, not all of whom appear explicitly in the text but who played an important role in my personal, as well as intellectual, development. I would like to thank all of them for the time and patience they took to share parts of their lives with me. This chapter is an elaboration of my undergraduate dissertation on corporate social responsibility, business and development, entitled 'The Value of Things and the Value of Actions: Political and Moral Encounters at Farmers' Markets in the UK', which won the 2011 Royal Anthropological Institute student essay prize.

To document the world of local food provision in this corner of England required a range of methodological approaches. Where conditions allowed, such as volunteer open days on farms and at the markets themselves, I engaged in participant observation, chatting informally with stallholders and farmers. This was combined with in-depth semi-structured interviews with farmers and food activists. The farmers were approached at the two farmers' markets in the town, but personal connections between them also brought me into contact with other farming realities outside the circuits of the markets. This broadened my understanding of the diversity of practices employed within the Sussex farming community, both organic and conventional. Finally, consumers were contacted and interviewed through the vegetables box schemes run by two market gardening outfits.

For a more historical background, there are some useful statistics on land-holding in nineteenth-century Sussex (Briault 1942; Godfrey 2002) that provide an idea of property relations in the countryside and ensuing social dynamics around work. Reed (1984) provides an interesting historical analysis, arguing for the resilience of a non-capitalist mode of production in nineteenth-century England, with particular reference to Sussex. Hughes (2011) provides visual representations of the countryside and its contribution to ideological constructions of meaning. For useful studies of farmers' markets, see Burr (1999) and Burr, Jewell and Rayner (1999), while Keenan (2012) provides an overview of local food in Sussex. There are also web-based sources on local food in Sussex (ARS n.d.), national initiatives (MLFW 2008) and food and campaigning in the UK (Sustainweb 2013). Finally, for further details on Transition Town Lewes, see the organisation's home page (www.transitiontownlewes.org).

8

Food Activism

We have called this book *Food for Change* because in it we meet people who are trying to resist the trajectory of the mainstream food provisioning chain and are seeking ways to operate outside it. The initiatives created by producers and consumers may appear to be traditional, or be revivals, but their substance is new. They are generated as reactions to the dominant model, both in their economic strategies and in their values. Since this is the essential context for our examples, it is worth reiterating the characteristics of the mainstream food chain.

The chain starts with the technological revolution in agriculture, which has transformed integrated farming systems and their organisation of labour. Agricultural production has been broken down into specific sectors and processes, each of which requires industrially produced inputs: machinery, diesel, plastic, chemicals and antibiotics. Labour requirements on the farm are dramatically reduced, since 'agriculture represents the increasingly residual activities which have resisted transformation into industrial processes' (Goodman et al 1987: 152). Farms have become more specialised and grown in size to achieve economies of scale.

A great deal of the foodstuffs leaving the farm gate goes to food processors and manufacturers. Often it is then broken down into component parts (sugars, starches, proteins, oils, fats) and recombined in various ways, together with other, industrially produced, additives. Most of the value is added at this stage, as maize is turned into breakfast cereal or tomatoes into pasta sauce. The length of the food chains and the disguised origins of the myriad ingredients in many of our everyday foods have led to repeated scares and scandals. Whether processed or not, up to 80 per cent of the food on our tables is bought from supermarket chains, where the global integration of markets becomes a visible reality. The scale of their operations and the possibility of sourcing food from anywhere in the world have exerted constant downward pressure on farm-gate prices. There are many interactions within this chain, but we will use the basic steps (production, distribution, consumption) as a framework to review the four case studies. Each section will bring out the way producers and consumers in the four locations engage practically with the economic pressures (and opportunities) of the open economy, and find ways to realise alternative values through some forms of closure. There are tensions and complex negotiations in all these initiatives, and we bring out some of them in a concluding section on the different political strands within them.

Before beginning, there are some general points which will clarify why people are trying to create different kinds of food provision, and the problems they face. Most of the farmers we met believe that conventional agriculture is not environmentally sustainable and that the mainstream exploits labour, paying very low wages throughout the chain, from the fields to the packing sheds and the supermarkets. They have strong political objections to both the environmental and social aspects of the system. These two factors also contribute to the low cost of food in the industrial North, a historic low in terms of average household budgets. So far, alternative food movements, even if we include the Europe-wide boom in grow-your-own initiatives, have struggled to make headway against the supermarkets. Various things could alter this balance, and the most far reaching of these may be changes in the mainstream. Conventional food could become more expensive, because wage levels increase or because industrial agriculture is forced to absorb the environmental costs of its production system (the degradation of soils and water supplies, for instance). Neither are very probable in the medium term. Rising prices for 'externalities', primarily the oil and gas which are essential to industrial agriculture, is more likely to shift the balance. It was a key reason for the transformation of Cuba's farming system in the face of embargoes, as they turned organics (or agro-ecology) into the mainstream, though this also required massive state investment and intervention.

In the industrial North there is growing critical awareness of the problems with industrial farming, food processing and corporate domination, a cultural change that nurtures and is stimulated by numerous initiatives in creating alternatives. At the present time, the farmers involved in these experiments have to live with one or both of the current outcomes: accepting low returns for their labour, or asking consumers to pay a price premium for the specialist foods they produce. This situation reflects the existing balance of forces between open and closed economic circuits. More specifically, it reflects the difficulties of creating more closed markets for food when that represents only one part of the economy, and raises questions about the kind of autonomy that can be created. We shall return to the politics of these issues at the end of the chapter.

Farmers and Production

The older generation of farmers in our three Mediterranean chapters grew up in a world where each household produced nearly all their daily needs, with a small surplus used to obtain other items through market exchange. This pattern, which by this point had disappeared in England, is sometimes referred to as the *campesino* model', or 'subsistence-plus'. Most farmers we spoke to simply say it was the traditional way, and for them it provides a recurring point of contrast to the direction taken by industrial agriculture. They make comparisons between the past and the present, and the conclusions are varied and subtle. There is some nostalgia for the old days, but it never becomes a version of 'we were poor but happy'. The Tuscan farmer who remarked that his farm used to feed a family of 14 is well aware of the kind of life this involved, but is still frustrated that the income they now get from the same land barely supports one middle-aged couple, despite all the technological investments. Reflections on what constitutes a reasonable livelihood merge with the other main theme in the comparison: the satisfaction to be had from farming work, its creativity or productivity, and above all the question of autonomy.

The Tuscan chapter gives the fullest description of the changes brought about by the farming revolution, though in some ways the situation there is rather anomalous. As share-croppers, the people described never had the same kind of autonomy as traditional peasantries, while after land reform, which introduced market-oriented production, they were still subject to many legal and operational controls over their farms. They eventually overthrew those controls, and were free to buy their own machinery, grow whatever they liked, and sell their farms, but it then became apparent that 'the market' had its own constraints, less visible, apparently less rigid, but in the end more powerful. One Tuscan farmer reflected, 'When the *mezzadria* finished I felt like a bird loosed from a cage, only to find myself in an aviary'.

Competing in the market came to require greater specialisation, growing reliance on expensive industrial inputs, and the end of a mixed farming system which fed the local population. In Tuscany we saw that for a period, investments, plus a variety of subsidies, welfare benefits and employment opportunities, did raise farm incomes. In the last 20 years this situation has gone into reverse as they suffer a classic price squeeze: the cost of everything they have to buy in order to produce has gone up far more rapidly than the price that can be demanded for what they produce. The squeeze created an outcome which was inconceivable previously; they could work all year and make a loss. It feels like a vicious circle generated by previous decisions. They abandoned crop rotation and spread ever more chemical fertilisers to increase yields; they bought in calves for fattening and raised them on imported feed. But on farms this size, in this climate, the margins are very tight. They are at the bottom of the food chain when it comes to buying diesel or chemicals, sectors dominated by corporations whose operations seem to them arbitrary and driven by speculation. When it comes to selling they find that, like the farmers in Sussex, they cannot set their own prices. Once upon a time, if bad weather in the district cut their yields, prices used to rise in response and there was some compensation. Now there are more globally integrated markets, and the sun is always shining somewhere.

In Andalusia, the opposition to industrial farming has its roots in a different historical experience. Here, there was no land reform. Instead, the great estates that dominated the region mechanised their agriculture, and hundreds of thousands of day-labourers lost their jobs. They emigrated to work in tourism and construction on the coast, or out of the region. Since then, new jobs have been created in agriculture, including intensive vegetable growing under plastic, but these mainly go to migrant labourers. This is all part of the astonishing boom and bust of the Andalusian economy over the last 20 years, but the crucial period for our informants was the dramatic rural crisis of the 1970s. Some were field labourers, others were smallholders at the margins of the estates; unlike those in Tuscany, they had not practised industrial farming methods, but they were all caught up by its effects on markets and employment.

This experience fed into existing left-wing political cultures, and meant that some responses to industrial farming systems were strongly anti-capitalist. The trade unionists and politicised workers opposed untouched concentrations of landed property and the corporations which dominated agro-industrial farming. After establishing their own cooperatives, some of them, most notably La Verde cooperative, continued to reject the industrial model because it represented another way in which capitalist enterprise would profit from their labour. Other values fed into their decision to farm organically, a concern with sustainability, with the health and flavour of food, including the importance of keeping local varieties rather than those grown from standardised commercial seeds. But the most striking theme running through their very eloquent judgments on industrial farming is that of autonomy. It is both an essential component of their version of anti-capitalist politics, and an aspiration which echoes through all the other case studies.

Few of the *néo-rurals* in the Tarn ever farmed using industrial methods, but they have been on the edge of it, and have very clear views on the disadvantages of what they call 'productivism'. It brings dependence on a chain of supplies over which they have no control. It also encourages specialisation and scaling-up production. Many of those interviewed had been tempted to do so as a way of increasing farm income but they have also found that they lost control of what mattered most to them: the farming skills of husbandry, responding to and shaping natural processes. The Sussex farmers expressed exactly the same thoughts: that good farming requires being independent, that 'you must stop expanding when it stops being a craft'. By contrast, depending on fertilisers is 'like painting by numbers'.

These farmers, who oppose the mainstream, have developed strategies to reduce their dependence and increase their autonomy by creating what we have termed a more closed economy. We outlined the general features of this in Chapter 1, and now we can see more clearly what it involves in practice. The first step is to minimise the money spent on farming inputs. Many have done this by creating mixed farming patterns which can reduce industrial inputs, since the waste from one part of the operation (straw, manure) is feed in another. Together with crop rotation, this is the major way of building up soil fertility. Some smallholdings practising horticulture do not carry livestock, but they tend to be located on deep fertile soil, and even so most of them find ways of getting manure. They keep seed for the next harvest, breed their own livestock and improvise machinery when they can. In the Tarn as in Sussex and all the locations, they practise the 'closed wallet' transactions mentioned by Ploeg, swapping materials and services with their neighbours without using money. In farming operations, they exchange labour and create work gangs for major tasks. They form cooperatives to pool resources and costs, from seeds banks to combine harvesters, though full blown production cooperatives have trouble reproducing themselves over time. In Andalusia and Tuscany, the dynamics of household demographics pulls them out of shape, so that after a generation the social capital is no longer held equitably.

The last strategy involves trying to produce foods with value-added, raising prices in order to increase incomes. This is more complex, and, in order to explain why, we shall also have to follow it through in the subsequent discussions of distribution and consumption. Not everybody we met is following this strategy. There are a handful of farmers, like the apple-grower in Sussex, who have a big enough operation to sell some of their crop to a supermarket chain. At the other end of the spectrum are those in the Tarn and Andalusia who would like not just to be an alternative to the supermarkets but replace them. La Verde, for example, can and does sell organic fruit and vegetables directly to customers at the same price as conventional produce in supermarkets. However, when they sell through the shops in the cities of Cadiz and Seville, there is a price mark-up (even if it does not return to the farmers) and a series of justifications for this, from organic certification to political solidarity. This is the more normal situation: farmers try to obtain higher prices by claiming that what was once traditional local fare (olive oil, tomatoes, cheese) is better than can be obtained in a supermarket, or constitutes a specialist product with limited production. There is confusion and overlap between these claims, from a vaguely defined concept of 'the local' to heritage vegetables and rare breeds, and to a range of certification schemes, such as organic, place-of-origin and Slow Food.

This becomes a highly contested terrain. The majority of these producers simply want to provide good fresh food to local people at a price they can afford. Some manage to do so; others end up, in the memorable words of one of our Sussex farmers, 'selling Christmas presents' or 'peasant food to posh people'. There are many reasons why it is more difficult for some producers to compete with the mainstream, including land prices, length of investment in the farm, or the local costs of housing and energy, and they all contribute to the difficulty of being 'an alternative'. These farmers want to be part of a different economy, one which is sustainable and rejects all the exploitative relations which underlie much of our cheap food policy. However, the mainstream conditions their actions at every turn, as the conclusion to the Sussex chapter reveals so vividly. They cannot escape the comparison with supermarket prices, which set the standard and often end up as the reference point for farmers when pricing their own goods. More than that, they get caught up in a whole set of conversations which takes place in terms of monetary value, rather than that of other values, whether sustainability or the anti-corporate strand which underlies much of the localism debate. And along the way, the farmers themselves are seen as contributing to an elite lifestyle, unavailable to the working class. Hinrichs (2000) discusses these issues in relation to farmers' markets and community supported agriculture in the United States. Her analysis of localised food systems reaches very similar conclusions, both about the way local 'embedded' markets are shaped by the mainstream, and the risks of them catering only to affluent consumers.

Certifying food as a way of achieving value-added is also contested. If we wanted a lively interview, all one of us had to do was bring three of these farmers together, ask them about the value of certifying, and let the tape recorder run. The point is that many of them are simply trying to farm in ways which they think of as traditional (or sustainable) and produce varieties of food which are indigenous, part of their culture and memory (or tasty, or nutritious). They are doing so in a world where those qualities are seen as exceptional. In their view, this is how food should be, and mass processed food is the problem. Why does something that should be normal need a label to prove it? We saw in Tuscany that in the world that the label tries to evoke, no labels are necessary. All this before the farmers start to talk about the bureaucratic control and costs of certification.

There are also disagreements about more fundamental economic issues. Small farmers seeking to add value through certifying their produce find themselves moving in much wider commercial circuits where everything can slip from their control, from the agencies which define what qualifies for the label to the organisation of markets. We have documented this throughout the book, particularly for organics, a sector dominated in Andalusia, as in California, by large commercial growers employing migrant labour, producing for sale in national and international markets. A similar process happens with place-of-origin legislation, and to an extent with the Slow Food certification scheme (Presidia), which valorise local specialities while being the antithesis of a local food economy. Commercial interests come into play because substantial profits can flow from this kind of recognition. The small step of trying to achieve a better livelihood in this way can undermine other objectives. The small farmers may end up rejoining a set of economic circuits they had tried to escape; the commitment to quality and authenticity is subsumed by the search for niche markets and higher prices, and the attempt to create a local food chain can become the start-up kit for internationally traded luxuries. No wonder the farmers argued so strenuously about the direction this road takes.

What is the economic outcome of all these production strategies for small farmers? One French farmer said that 'wealth is elsewhere', a judgement echoed in other locations. There are a handful of farmers supplying local markets who are reasonably prosperous, and there are some who own property which would be worth a great deal if they sold up. Land prices are very high compared to the income that can be generated from them, especially in Sussex and Tuscany. But overall, the farmers we met had very low cash incomes, less than €1,200 a month, and this is confirmed by statistical material on British farm incomes, from Ploeg's research, and elsewhere. In an earlier study, Jeff Pratt analysed Tuscan farm incomes, using the unusual wealth of data which was in the public domain (land holding, family composition) or easily obtainable from other sources (crop prices, employment records and so forth). Direct questioning was not usually appropriate because there are many things that households chose not to tell you: the pension income from grandmother, the farm subsidies (which are complex and variable between countries and regions). There are also many things they do not calculate or which make comparisons difficult, and this in a different way reveals a good deal about a small-farm economy. Unlike the urban self-employed, they rarely have rent or mortgage payments on their homes, whether that is a farmhouse or a caravan. They produce much of their own food and fuel. Their cash income (from sales, off-farm work, welfare benefits) fluctuates dramatically from month to month and year to year, and is spent on unavoidable expenses on the farm, new investments and consumer items. The proportion spent on each of these categories is decided by urgency and pragmatism. Taken all together, this means that most of them do not know how much they earn in cash terms, and that to some extent the question does not make sense. We also saw that our farmers did not know what it cost to produce a lettuce or a chicken, and that an accountant's calculations did not make much sense either.

At some level those we talked to have chosen to be farmers, whether *néo-rurals* or the survivors of the rural exodus. When they are invited to reflect on their reasons for choosing to farm, they talk mainly about the satisfaction which comes from this kind of production, about autonomy, and about a range of social and political commitments. They do complain about money because they feel people do not understand the long hours they work and the expenses involved in some forms of farming. But they are not 'in it for the money', and do not measure their success by the kind of consumption-based prosperity which dominates the societies around them. It was striking how many used the national minimum wage, or its equivalent, as a benchmark for what they thought was acceptable, even calculating back from it in setting their prices. This opens up again the question of the just price, and all the other issues which emerged in the discussion of a closed economy, and we shall return to these themes in the next chapter.

Distribution

It is very hard to organise distribution networks for this kind of small-scale farming; the logistics are tough. In the background, shaping everything, are the supermarkets, with their highly efficient monitoring, stocking and transport systems, bringing a vast range of products from all over the world into each store. They set the lowest overall prices for a 'weekly shop' – though on any one day some of their food will be more expensive than in other outlets. They also sell a growing range of non-food items, from toothpaste to clothing and books, so that almost all the consumers in our case studies (and the authors themselves) go to them as part of a weekly or monthly pattern of shopping. This gives rise to a kind of dual pattern of food provision which we shall return to below.

In constructing their own food chains, these farmers have relied heavily on direct sales using a range of methods – farm shops, farmers' markets and box schemes in particular. In the Tarn there is a version of community-supported agriculture, while in Andalusia there has been a move into local government contracts, supplying schools, and a version of 'solidarity economics' in the production–consumption cooperatives established in Seville, Cordoba, Granada and other cities. Many of our informants were part of international organisations and networks which pool information about running and expanding these schemes. Direct sales are the simplest and most effective way of increasing income. Farmers in Tuscany making oil and wine are effusive about the advantages of sales to locals and tourists compared to the struggle to enter the commercial circuits of restaurants and boutique shops. They have to tour round towns and trade fairs, leaving freebies, negotiating substantial discounts, facing late payments and fraud. In Andalusia, the Tarn and with some of our Sussex producers, we heard a much more political objection to retail systems, which leave so much profit in the hands of middlemen and so little return to the farmer.

Direct sales also have their disadvantages: primarily, that they take time away from farm work. La Verde cooperative reluctantly shut its farm shop because of the need to stop digging in order to sell a few pounds of vegetables to passers-by. Farmers' markets require late evenings loading up a van on top of the long day running a stall. They may enjoy the contact and feedback between producers and consumers, and there are long-term advantages in this visibility, but it creates pressures which farmers commented on in all the locations. Box schemes undoubtedly work, and provide a substantial proportion of many farmers' incomes, but there were regular comments about limits to further expansion. In Sussex it was partly because, amongst professional middle-class families, a key 'demographic', there has been a recent expansion in allotments and grow-your-own schemes. In the

Mediterranean, many households are at most a generation away from rural life and have access to a patch of land. In Spain, even in the cities there are radical political movements taking over peripheral land and creating urban gardens.

What proportion and range of consumers' food needs are these alternative food chains providing? Sussex is clearly the odd one out, since even with the variety provided by mixed farming systems and niche producers, we are a long way from a local diet. By contrast, in the Sierra de Cadiz, only fish, coffee and bananas cannot be sourced locally, and a similar picture holds in the Tarn and Tuscany. However, there are always difficulties around providing variety, both for box schemes and farmers' markets. If an individual farmer tries to provide a weekly box of all the vegetables a family consume, over the year they have to grow a very large range of crops. This makes it difficult to get the economies of scale achieved by commercial growers, something which the Sussex farmers talked about a good deal. One way forward is collaboration, so that each farmer produces fewer crops, which are then pooled. Riverford Farm in Devon (now a franchise) is a more formal and commercially oriented example of this. It is a mutual cooperative of a group of farms, which allows for greater specialisation and mechanisation in growing, and a depot which sends out 45,000 boxes of organic vegetables a week. In Sussex and in the Tarn we find a more informal system of sharing seeds and surpluses. In Andalusia, La Verde itself pools the products of a dozen producers, and was once linked into a regional network of organic growers who evened out shortages and surpluses and shared specialist crops like oranges.

If these farmers value their autonomy above all, it is not surprising that collaboration can be difficult. At the most pessimistic end of the spectrum we heard this quip in Sussex: 'How do you get three farmers to agree? You shoot two of them'. In many cases there are political movements creating collaborative networks, but most farmers' markets have been established by municipal governments, who have tried to create a critical mass of sellers and a range of products, sometimes regulating to avoid too many in any one field. In this sense, the Tarn examples are clearly a success story, developing out of a particular food culture and enabling its continuation. They have become a one-stop shop for food, with everything available and locally sourced: bread, cheese, fruit, vegetables, wine, meat and processed goods. They are held several times a week, which means customers can routinely shop there. Without this range and regularity, they cannot aspire to be much more than a place where people supplement their supermarket shopping with a few specialties.

Consumers and Consumption

The farmers' livelihoods and their distribution networks presuppose the existence of shoppers looking to buy food with certain qualities. These people as consumers are drawn towards products and farming systems they see as threatened by mainstream production and retailing. In that sense, their purchasing decisions are key to the whole business of 'food for change'. This means we need to look at what they want from the products they buy, and understand the ethics behind their behaviour. In doing this we should recognise, as many of the interviewees do, that people are not always consistent in their actions and motivations. Dealing with the complexities of shopping is best accomplished by returning to the case studies.

Before reviewing the material, however, we need to return to arguments about the place of consumption in contemporary life. Two themes stand out. First is the increasingly attenuated relation between production and consumption, with the retailing and marketing industries operating in, and profiting from, the ground that opens up in between. Thus Miller (1995) describes the overriding experience of modernity as a sense of rupture – a process in which people have ever less idea of the conditions under which the things they use are made. Alongside this is a second characteristic: the process by which people construct identities through their individual choices. So consumption comes to be understood as a project by which we, as individual persons, engage in the enterprise of creating our self-identity through the choices we make.

These two overlapping strands, rupture and self-identity, come together in interesting ways in relation to food. In Chapter 3 we argued that food is a privileged domain in which people try to counteract the world without boundaries generated by open markets. We also argued more generally that the separation of consumption and production into distinct domains is problematic, and part of the mystification of capitalism. In effect, drawing production and consumption towards one another is an attempt to escape rupture and find more meaningful relationships than those denoted by alienated consumer identities. This is what is achieved, or at least sought, through closure. Food is good for this because it provides a metaphor for, and sustains, clearly demarcated and hence closed systems at many levels. But the contradiction people must deal with is that the means to make this escape is precisely through more shopping. How does this work at the level of everyday practice? To answer that question means looking at political cultures around food and shopping, drawing particularly on the English and Spanish cases.

Lewes, which is at the centre of our English case study, has a long history as a market town. Retail is still important, with a strong tradition of local, independent shops, while the importance of local markets is exemplified by the Lewes pound, which existed for most of the nineteenth century, and was revived in 2008. The benefits of the local currency are said to be economic, social and environmental: to multiply wealth and build resilience in the local economy, to support local businesses and strengthen relations between shops and shoppers, and to reduce carbon footprints (TLP 2013). Such characteristics indicate a culture of consumption very different from the 'clone towns' dominated by chain stores, out-of-town shopping and deserted town centres. On the other hand, the presence of three national supermarkets suggests a more complex pattern of consumption than rhetoric suggests, not least in socio-economic terms. There are wealthy professionals living in leafy suburbs, serviced by a high-end national supermarket, delicatessens selling specialist foodstuffs, and the farmers' markets. There are also substantial working-class housing estates, home to white- and blue-collar workers. On the southern edge of the town the supermarkets compete for more price-conscious shoppers' attention. The participants in the box scheme discussed in the English case study tended to fall somewhere between the two groups; that is, they were generally professional people on middle incomes. In the words of one of them, they enjoy 'educational rather than economic capital'.

This is reflected in shopping and food provision strategies. In a series of interviews, we asked participants in a vegetable box scheme run by a local grower about their shopping and eating habits. The overwhelming impression was that the goods in the boxes were special, but in specific ways. People enjoyed the element of surprise, of not knowing what the contents would be, and trying out new foods and dishes. This aspect makes a virtue of bucking the trend towards consumer choice; people actively relished constraint in what they cooked and ate. The limited range was further attractive because it was underpinned by ideas about seasonality – making do with what the local ecology could produce at any given time of year. Further, people said cooking using basic ingredients they would not normally buy was pleasurable and social, especially when children became involved in food preparation and eating was a family event.

Most of the interviewees also supplemented what they bought from local suppliers by growing fruit and vegetables on allotments and in their gardens, as well as picking fruit from the countryside. Foods procured in this way were distinctive in the same way the vegetables from the box scheme were special. Both were described as an antidote to ready-made meals and TV dinners, and justified any extra cost or effort in producing them. There appears to be some confusion at times about the boundaries between these categories. One market seller recounted how he spent ten minutes discussing the merits of local apples, and was taken aback when the shopper then asked if he sold avocadoes. Most people also bought cheap bulk items like pasta or tins of tomatoes and ready meals in the budget supermarkets, though this was described as atypical and a departure from preferred shopping habits. Overall, the evidence is that the people in the box schemes are inclined to pay more for locally produced fruit and vegetables, exotic and higher quality processed goods, and this expenditure testifies to the quality of the food as well as becoming proof of a political commitment. However, it is worth emphasising this is a sample of more middle-class, professional and educated, though by no means wealthy people.

In the small town at the centre of the Spanish study we find a different set of relations and food practices. In a parallel set of interviews, housewives were asked about their shopping habits and food procurement strategies. In Spain the interviewees are much more connected to the countryside; often they have land or family and friends who farm, and who supply them with a great variety of food, including meat, eggs, honey, fruit and vegetables. Although the general sentiment is that such foods are extraordinary in terms of taste and freshness, there is no expectation that this would make them more expensive – quite the reverse. Local foods are special because they build resilience into the *pueblo* in an area where supplies from further afield have historically seldom been reliable or affordable. These are people who knew about food sovereignty through necessity, long before it became a buzz word.

It is difficult to gauge the day-to-day importance of these local channels of food supply in Spanish towns. However, there is no doubt that people enjoyed talking about them, and they expected locally produced food to be competitively priced. In other words, the fruit from an uncle's tree or their neighbours' carrots formed part of provision strategies, helping to reproduce the household through saving and recycling. They constituted part of a local economy in a much more profound way than in the English case. Producers expected to sell their produce locally because this was part of their political culture; they do so either directly to the consumer or on the street or through shops. Hawkers walk the streets or set up impromptu stalls selling wild asparagus or cactus fruit, and there are local laws prohibiting outsiders from doing this. In the town marketplace, the stall that sells local fresh fruit and vegetables has comparable and sometimes cheaper prices and longer queues than the seller opposite whose produce comes from further afield. Meanwhile consumers expressed a preference for locally-owned supermarkets, which were more like a traditional English grocers, than national chains. So, in the case of unprocessed food, people from across the class spectrum turn by preference to family, friends and local suppliers before supermarkets.

This is not to deny the growing strength of national supermarket chains. People drive to stores to buy soap powder, bulk processed food and items unavailable in the villages, but there is a sense that these are special trips to purchase unusual goods, with people often dressing up for the excursion. The high proportion of foreign shoppers at these establishments in an area with little tourism seems to reinforce the extraordinary, modern, international flavour of expeditions to large supermarkets. In that sense, these excursions are different from everyday shopping and are not always or necessarily about saving money. Conversely, local food is considered ordinary in the way supermarket shopping is in England.

It needs to be emphasised, however, that we are here talking about small towns. In Andalusian cities, food from named producers is sold in specialist stores with premiums attached, much as in England. The specialism may be politically marked and sold through a cooperative, or it may play on ideas about health, but it is seen as different from normal supermarket goods because it evokes a closed economy. Labelling exemplifies the difference between these two worlds. In small towns, organic food is confounded with local food and anything grown locally is considered 'organic', whether it has a label or not. Indeed, this conviction is so strong it seems to overrule the question of whether chemicals have been applied during production. A retailer who sells certified organic goods in one small town lamented that 'people just don't understand the concept'. By contrast, in the cities the organic or fair trade label is a prerequisite.

In our case studies we have a range of attitudes towards different kinds of consumption. The French case seems to lie closer to the Spanish example, with closer links to the countryside and the expectation that growers should provide normal, everyday food to everyone, and at affordable prices. In England and Italy there is a greater sense that there is a market for locally produced, organic or other quality food, and the expectation that this carries a price premium.

However, these differences are perhaps more real than apparent; consumption patterns may be more a reflection of the degree of urbanisation and the distance from closed systems of food production than anything else. If you live and shop in an open economy where the conditions of production are largely invisible, then when they do become visible the food is seen as special and can command a price premium. On one level, in terms of the consumption of modernity, there is a consumption of nostalgia by those who can afford it. Where does that leave the practice of ethical consumption as a way of opposing cheap mainstream food with its hidden costs? It is to that political question we now turn.

Politics

The alternative food movements we have documented embody an array of political ambitions. Farmers and agricultural labourers in the four regions are heirs to a variety of political traditions, which they have also renewed and modified through their own activism. Their livelihood strategies have created more direct links to customers, many of whom have their own political purposes which they try to realise through ethical consumption. This relationship can take a more concrete form in the organisation of cooperatives and networks linking producers and consumers, or through the ambition to realise a local economy. In this concluding section we will outline some of the key issues in this complex political landscape.

In Tuscany and Andalusia we met farmers whose activism began in a period of class politics. As share-croppers or labourers they worked within movements which challenged property rights, forged alliances with other workers, and sang songs in the future tense about freedom and the end of exploitation. They achieved some partial successes, which is why they are farming now, but this kind of collective and transformative ambition, framed by an understanding of unfolding world history, seems to belong to another era. Today we are more likely to encounter a narrative about sustainability which portrays the future in negative terms as increasing crisis and degradation. This has its more radical edge in the Tarn 'de-growth' movement, and its more comfortable form in an English Transition Town. It is a complicated political landscape, partly because food movements can fit with such a variety of wider projects.

Many writers on these movements make a contrast between alternative and oppositional objectives, but it is not an easy line to draw, and would in any case be rejected by those in the Tarn or Andalusia, who are close to anarchist thought and practice. All we can do here is signal a few themes, especially around individual and collective action, and the issue of autonomy. The implicit comparison is with those earlier periods of political action which involved a class struggle between two socially and geographically located forces. The comparison is not invoked as a paradigm of what politics should be, but as a way of signalling how much both the strategies and the forms of mobilisation have changed.

The farmers in our case studies, whatever their historical roots, are now all self-employed, and their struggle for autonomy is against 'market forces', behind which stand large corporate interests. One strategic response has been the formation of production cooperatives, found for example in Tuscany and Andalusia, though in both cases they have encountered difficulties renewing themselves after a generation. A more widespread strategy for increasing autonomy is through enlisting the solidarity of consumers. This introduces an appeal to specific moral or political values in the marketplace, and in the sphere of exchange more generally. This can play out in various ways, some of which are more individualistic, and some more collective in their organisation. Ethical consumption has a long history, from boycotting oppressive regimes to today's fair trade movement (Carrier & Luetchford 2012). In the food movements, one common ethical principle is environmentalism, and specifically sustainability, an issue which is widely portrayed as a universal problem affecting everyone on the planet, and one that can be approached through the model of individual lifestyle choices: use the bike, turn down the heating, eat local. Without denying that such action can have results, critics of ethical consumption argue that it stays within the neoliberal paradigm of individual choice in the market place, and deflects attention from the structural processes which create the 'unethical' reality. Slavoj Zizek has gone further to argue that corporations from Starbucks to airlines have incorporated the appeal to an ethical bonus into their marketing strategies, in what he calls 'cultural capitalism' (Zizek 2009; see also RSA 2010).

Solidarity between producers and consumers often involves some measure of direct contact. Farmers like to establish a personal rapport with customers, though many expressed frustration at what they felt was an 'unequal' commitment to their shared aims, combined with occasional resentment that 'the customer is always right', or at least can call the shots. In the Tarn and in Andalusia we found moves towards a more structured long-term relationship, not just through box schemes and farmers' markets, but through community-supported agriculture and solidarity economics. The hope is to go beyond a relationship between groups defined as 'producers' and 'consumers' to a shared enterprise within a larger framework. In the most radical experiments, producer–consumer cooperatives try to collapse the distinction. Slow Food has adopted a similar aspiration with the notion of 'co-producers', without moving much beyond feel-good rhetoric and the model of individual choice. Where a more long-term and collective framework does emerge, it can be based on the network form, or on locality. In both the Tarn and Andalusia there are loose networks of political activists of the kind that has become common in the 'alter-globalisation movement (Maeckelbergh 2009). Alongside farmers and local supporters, activists include students, academics and trade unionists in neighbouring towns and cities, people who are opposed to unfettered capitalism and consumerism, and are committed to developing more egalitarian and democratic forms of economic organisation. These networks are often named branches of regional and national organisations: Confédération Paysanne and Nature et Progres in the Tarn, and the Federacion Andaluza de Consumidores y Productores Ecologicos (FACPE) in Andalusia. They in turn may be affiliated to international organisations, such as the Via Campesina, which campaigns for food sovereignty and against the WTO. These larger federations are important in sharing information about the creation of economic initiatives, while their books and videos widen the range of people with critical knowledge about farming practices and the global food trade. Nevertheless, the basic shape is not hierarchical, and the international organisations do not set the political agendas or the activities of smaller-scale networks. In effect these are umbrella organisations for initiatives whose dynamism and success depend on grassroots support and conditions.

Alongside these networks there is also an attempt to create solidarity at the level of the local, an ambition which is a central element of many food movements, partly because of the link with sustainability. In Sussex the local is a fuzzy concept, a category of people who live in the same town or district, whose lives take them in many directions, many to professional jobs in London. So far, they share the location of their homes but little else in the way of a local economy. The farmers there say that the attempt to create local food chains, while laudable and valuable, is still a fraction of what is needed. Turning the perspective around, (some) people may value (some) closure in terms of their food supplies, but they live the rest of their lives thoroughly immersed in the open economy. This is a recurring issue, even where localities are much more clearly defined and have a recent history of substantial economic autonomy. In Andalusia, the *pueblo* has in the past represented a much more bounded economy, and in the district we studied is still a much more homogeneous entity. But even here people do not live by bread alone: they also buy electricity, cement, books and computers. The farmers – frugal, inventive, committed to sustainable local food supplies and autonomy – do not often reflect on themselves as consumers as well as producers. This is not a criticism of the farmers: they do what they do, grow food, in a society whose older generation can remember hunger. The issue is that they and their supporters in these solidarity networks need other goods in their lives, and the way these are produced and exchanged does impact on food circuits. The 'just price' for a basket of fruit depends on what it can be exchanged for, and from that perspective the long-term outcome of these initiatives will depend on transformations in the wider economy.

There is another way: going off the grid, and off the map. In the Tarn, as in the Larzac (Williams 2008b) and elsewhere in Europe, some are attempting to create autarky, exploring ways of producing all they need: food, clothing, housing and energy. They reject capitalism, consumption and the paradigm of growth as progress. Some reject money completely, since they think it is incompatible with a society which is sustainable and free of exploitation (Nelson & Timmerman 2011). What was once for peasants a necessity and a defence becomes a morally charged search for coherence, and is often informed by spiritual values, as it was for earlier 'back-to-nature' and other counter-cultural movements. Every step to greater autarky is an attempt to rethink the realm of the necessary and the noise of false needs. The corollary is that those who achieve this most completely are also those who risk isolating themselves most completely from society and the forces they seek to resist. Williams (2008b) explores very subtly the political and cultural implications of this attempt to maximise autonomy; in the last part of this book we shall reflect on the economic implications of being inside and outside these circuits.

9

Economics and Morality

I n this chapter we will broaden the agenda. We have looked at
the practical issues and difficulties faced by those constructing
alternative food movements, and we suggested that the opposition
between open and closed economies can make sense of their strategies.
Now we want to look more generally at what this opposition involves.
In analysing what is at stake in the various case studies, we draw out
the importance of the values embedded in economic activity and in
the social relations that activity creates. This also means disentangling
the significance of money and monetised value for activists in these
movements. In exploring these themes we shall draw on economic
anthropology, which has developed both a series of reflections on
the relationship between money and morality, and also developed
concepts which illuminate the distinction between open and closed
systems. The final section widens the agenda even further by pointing
out how the values, distinctions and tensions found in relation to
producing and consuming food are also present in many other parts
of our everyday lives.

We chose the terms open and closed economies to refer to the most
general aspect of the case studies because they seemed closest to our
informants' own understanding. Sometimes the distinction between
open and closed emerged in relation to the actual movement of goods
and services. Farmers are very conscious of everything that arrives
or leaves through the farm gate; it is a very practical awareness built
into their everyday lives because it is a central part of their struggle
for autonomy. The creation of a closed system is also an explicit
objective of organic farming. We find an underlying move towards
closure when farmers and consumers form longer-term alliances,
either through cooperative networks or in their move towards a local
economy. Even in less structured situations, where consumers chose
certain kinds of foodstuffs for those meals which are important to

establishing shared identities and memories, we find the same theme, if in a more attenuated or metaphorical form. The significance of a closed system and its values emerges in contrast to another economic world, that of industrial farming, supermarkets and mass-produced food of unknown provenance. There is often a defensive tone to these movements because the open economy is experienced as a constantly destabilising force, sucking out value and reducing autonomy. We would stress that strategies of closure are always shaped by the wider open economy, and that we view closure as relative, an ambition which orients practice, with considerable ambivalence about what absolute closure would entail.

Self-Sufficiency … and Mutuality

In summarising the way farmers develop a more closed economic system, we have stressed, like Ploeg (2009), the strategies of self-sufficiency. Some had never moved far from older peasant practice, some had tried the open system and failed, and for some it was a choice generated out of a wider political culture. For all of them, farming is an endless engagement with nature, with a particular patch of land, its soils, slopes, weather patterns and pests. Producing any food is a struggle: making the farm produce more or better quality food requires skill, experience and the capacity to experiment. It requires constant watchfulness and flexibility: work carries on late when the weather is about to break or a calf is due. It requires long-term commitments. Both in Sussex and Andalusia, market gardeners said it had taken seven or eight years to build up soil fertility, and the same applies to building up a flock of sheep or seeing an orchard into mature production. The ambition is not just to produce food this year, but to do so in a way which enables you to produce food in the future. Nobody is 'in it for the short term', and farmers need a particular kind of patience and forbearance to accept the levels of uncertainty about the outcome of their work that farming generates. These are very different attitudes from those required by most kinds of wage work, or the kind of industrial agriculture which is closer to 'painting by numbers'. So there is more to all this than material processes and a kind of stubborn individualism. It is also a question of skills and values.

There is a great deal of creativity in this kind of farming, if not that of an artist (though there is an aesthetic dimension) then at least that of an artisan. This creativity is valued in itself and generates value, though not in the first instance that of the marketplace. Spelling out directly what is involved in this value is not easy – much is unspoken and has to be deduced from what people say they are *not* doing, and from the way the activity is embedded in social relations. So we have to take a roundabout way to the issue, and some aspects will only become clearer after a discussion of money and markets.

Obviously there is a strong personal pride in skill and creativity, as we saw in the interviews. It is combined with the hope that the importance of the work will be recognised or 'validated' by others. One way this is achieved is through direct sales. There is personal contact between producer and consumer, and even where that is missing there is the consumer's knowledge of where, how and by whom the food was produced. This connection to the details of production are at the root of claims about food quality and authenticity (Pratt 2007), and are an obvious contrast with mass production and processing. When Enrique at La Verde in Andalusia says, 'We use the name and our prestige as a human group to sell', there is the underlying assumption that the connection between the producer and the commodity continues to be a live presence, like the spirit of the gift, in the marketplace. The food sold is not the outcome of undifferentiated labour, but incorporates the work and creativity of particular people. We have emphasised how direct sales increase livelihoods, but they also provide a bounded arena within which the value of this kind of work and creativity is realised.

The importance of the social context of these values emerges in other ways. We have referred a good deal to the concept of sustainability and its importance in developing closed farming systems. Most of the discussion has been about the interaction between human labour and nature. This was a major theme at La Verde cooperative in Andalusia, whose founders talked about techniques for building up soil fertility, the creation of biodiversity, their pride in planting 10,000 trees and the effect of this on the micro-climate. They insisted that the value of this effort 'cannot be measured in pesetas, since all you put in over many years is your energy, your hands, your imagination … [T]his is sustainable development, this is wealth creation'. And at the same point in the interview they added, 'This is our effort on behalf of future generations'. In other words, all this activity is not only an engagement with nature or about personal satisfaction; it is part of a long-term relationship with a group of people. It is in the same spirit as the Tuscan farmers who direct their energies into passing on a viable farm, and a way of life, to their children. In the production cooperatives there is a very similar concern for the next generation, though it is also framed within a wider solidarity with those who share the same ambitions.

These patterns of social relations can be thought of as spheres of mutuality, to use a phrase from Gudeman (2008). They include the multi-generations of a farming household, a production cooperative or a solidarity network. The term is loose enough to cover everything from highly bounded kinship groups to notional communities, and also indicates that these spheres can overlap. In practice, the sharing or distribution of whatever is produced constitutes the definition and boundary of each realm. It is another way of thinking about the importance attached to autonomy, and about the widespread concern to prevent the value of work and creativity being appropriated by others. The significance and moral qualities of mutuality usually emerge when people contrast them with other kinds of economic activity, premised for example on competition and a quantitative and atomistic conception of self-interest. Jordy in the Tarn talks about work and sharing, and how in order to establish cooperative projects 'we have to get out of this calculative thought that is polluting our minds', to trust the collective rather than profit seeking.

The kinds of distinctions articulated by our informants are found widely within economic anthropology. 'Calculative reason' is Gudeman's version of Jordy's term, and we shall discuss it more fully below. In turn, this is a version of Weber's formal rationality, which has a long pedigree in debates about the principles of economic activity. Jordy's longer reflections on exchange echo the distinction between what Sahlins called the generalised reciprocity of sharing and solidarity, as opposed to balanced reciprocity, which still involves counting (Sahlins 1974: 193–95). In his epic study of debt, war and peace, Graeber (2011) uses the concept of communism to express a similar notion of mutuality. It is the contrast between two kinds of economic activity which runs through all this ethnographic material (and anthropological analysis), but there are significant differences in the way they are constructed. For those, like Jordy, seeking autarky, the whole world of money and markets connotes egotism. For other farmers, the principle problem is corporate power. To grow conventional food they become dependent on Monsanto, and to sell it they become dependent on a supermarket chain. For them, the 'market' signifies not calculative reason or competition but monopoly control.

Money and the Just Price

This brings us inevitably to the question of money. It is one of the trickiest themes to summarise, partly because there are many substantial discussions of money in contemporary anthropology (e.g. Hart 1986; Graeber 2011; Gudeman 2008; Parry & Bloch 1989). They have revealed the different origins of money, and its multiple uses: the number of uses varies in each author, but normally includes money as measure, as medium of exchange, and as a value in itself. Parry and Bloch (1989) have analysed a variety of symbolic representations of money, and argue that Western moral ideas, associating money with depersonalised relationships and self-interest, is far from universal. Instead, they suggest a more general opposition between two 'transactional orders': one of short-term individualistic competition, and the other of long-term social order (ibid.: 24). We shall draw on these discussions to open up some curious tensions in this ethnography, tensions which take a form which is not universal, but which still has a much wider relevance.

Money certainly has many uses, one of which is to allow people to exchange goods and services. Exchange can take place without money, but its use is not intrinsically contrary to the practice of mutuality, trust and equity. Yet often in the interviews we hear an assumption that money connotes activity which is emptied out of humanity and any moral purpose. Money invokes markets, and market relations suggest the egotistical pursuit of personal gain or the unfettered power of corporations. What is more, using money for exchange implies that everything has, or might have, a monetary value; thus everything is commensurable in monetary terms, which drives out other values. These slippages are found regularly in the interviews, and some may still be there in our summaries. For some people we met, the love of money is the root of all evil – as it has long been in one strand of anarchist (as well as Christian) thought. But there is a more widespread wariness about what happens when any kind of money calculation enters into an activity.

Here are two very different examples, taken from earlier chapters. The first is from Loreno, the organiser of the production cooperative in Tuscany, commenting on a neighbour who had left the group to work as a labourer.

> I could not do that. Man's initiative finishes that way, you become an object, it is no longer you who decide, who creates projects, who makes plans, who develops your own personality or your will. For a lifetime you have to do what you are told. You sell your labour to others. I do not feel like working like that, as if the principal thing in life is getting money, taking the attitude that I don't care whether the job is done well or badly, they pay you all the same.

The comments start with a theme which is by now familiar: the commitment to personal autonomy, creativity and the long term. It then shifts (in a way which not everybody shares) to the view that if you take paid employment then pride in work disappears, along with any sense that work should be creative or produce something useful. The introduction of money has squeezed out other values.

The second example is from Daniel in the Tarn, explaining why he downsized to escape the commercial mindset which had come to dominate his daily farming practice and his dreams.

> You get to the point where you think: I am fed up seeing my lambs represented merely as bank notes. On the one hand, you are in a world of finance, and on the other, you live and work with 'living matter'. At night I had nightmares where I could see bank notes coming out of the arses of my ewes. Do you realise the contradiction? What is a 10,000 franc note? You can set light to it, it is not worth anything! And you have a living animal, who goes through the effort of birth, and this is worth a mere 10,000?

Here is the other end of the chain: all the productive work with nature assisting new life is perverted when directed towards maximising the financial returns on a commodity. Money is shit.

Daniel has moved from commercial sheep farming to a more integrated closed system, but he does still have to sell his lambs. It may be a regular event, yet it contains all the incongruity he describes. What connection is there between a live animal and a banknote, between something with unique qualities which he has nurtured and a quantity which makes everything commensurate? There are a variety of reactions to this equation. One is autarky: don't buy or sell anything. In practice, no European peasant in living memory has been able to achieve this completely; there has always been a minimal recourse to markets for goods which cannot be produced locally. However, autarky is always there, as an ideal for a few, and for many others a kind of yardstick to measure their own escape from the circuits of the open economy. How much of the stuff we need can we make ourselves?

Another reaction is to try to inject some moral criteria into the exchange. In capitalist societies there are polarised attitudes towards competitive markets: some see them as the source of collective progress, well-being and liberty, others as an egotistical moral-free zone. Mostly in our interviews we encountered the second view: markets are driven by the self-interest of buyers and sellers, while further back in the process of price formation there are other factors which are contingent or arbitrary. So the equation between a lamb and money should have some other dimension: the money should allow the farmer to complete an exchange which is fair, one that enables them to buy something equivalent. This is the just price, but giving reality to this ambition is very difficult. We noted in the previous chapter that a very common approach was for the farmers described to set their prices so that over the year income approximates that of a person on the minimum wage. This is very rough and ready, but calculating hourly rates for labour is not realistic.

The 'just price' in alternative food systems does not of course exist in a vacuum. The mainstream market price is inescapable, both in terms of economics and in shaping people's understanding and attitudes. In some cases, these farmers can match supermarket prices even when producing organically, in others they cannot, especially in England. As a result, the farmers have to ask something extra for their food, a price premium, and find customers who want to pay it. The price premium is discussed in at least three ways, the first two of which also represent ways in which moral principles are developed within economic relations. It can be represented as a political commitment to support the livelihoods of small farmers, to support sustainable agriculture, or to boycott the mainstream because it is associated with corporate power and exploitation. This overlaps with a second strand, creating a market which has some closure from the open economy. We have discussed this in terms of spheres of mutuality, whether in the construction of a local economy or the networks of solidarity economics. The third strand is that the price premium is itself justified in purely financial terms; in other words, it delivers quality which is worth paying for. This can be on the grounds that the food is fresher because of shorter supply lines, or more healthy and nutritious (especially with organics), or that it is a local speciality which is not otherwise available. We have discussed these discourses about quality at various points in the book, how they link production and consumption, and how they move into arguments about monetary value.

Realms and Values

These comments on the just price bring out the dynamic tension between open and closed economies. At one level, closure is the attempt to create a space within which unfettered markets and calculative reason cannot operate. At the same time, the strategies, the success and the significance of closure are shaped by the wider economy. We need both perspectives. A historical comparison should make this clearer. In the Tarn and in Andalusia we can find a few households farming in pretty much the same way as their predecessors 50 years ago. They practise mixed agriculture, with traditional breeds and plant varieties and few industrial inputs, in order to produce food for themselves and a local population. Their predecessors were peasants, and it was the only kind of farming around. Our contemporary households have chosen this path, creating a more closed economy in a rural world transformed by new technologies and global markets, and their strategies have been shaped by this wider economy. So although farming practice and values are very similar in both periods, the context is very different, and if we lose sight of this we cannot understand these farmers' economic choices or the political and cultural significance of the food they produce.

There is another aspect of the relationship between open and closed economies. Capitalist enterprises are constantly trying to encroach on the economic activities generated within realms of mutuality, and, in the process of turning them into marketable commodities, appropriate the values associated with them. In the case of food production and consumption, traders, manufacturers and supermarkets try to capture the value produced outside the mainstream. Sometimes this is a straightforward economic appropriation, as seen in one strand of the development of organic farming or in many certification schemes for local specialities and artisan foodstuffs. Sometimes this is a cultural appropriation via the advertising industry, so that mainstream products are layered with the values of the domestic, the traditional and the authentic: cakes baked by grandma in her kitchen, pasta made from wheat harvested by farmers with sickles in their hands and straw in their hair. It is hard to know whether this sales strategy is premised on customer ignorance or postmodern irony, but clearly the appeal to nostalgia works at some level and has become more prominent. The relationship between open and closed economic systems is obviously not symmetrical, so how should we understand it?

At this point it is helpful to return to related work in economic anthropology. One writer we have already referred to is Stephen Gudeman. In Chapter 2 we discussed the distinction between house and corporation, developed from an analysis of Colombian rural society (Gudeman & Rivera 1990). More recently, Gudeman has taken the house model as just one example of 'a communal or mutual economy' (Gudeman 2008: 4), which is found all around the world, and everywhere coexists with a sphere of competitive trade. Like other anthropologists, Gudeman insists on a very broad definition of the economy, wider than the conventional wisdom that concentrates on the activities and relationships which involve money and can thus be quantified. There are so many other ways in which we produce things, 'services', knowledge and art, and there are so many other ways in which we share or allocate them. Of course, in our society, cooking a meal, minding children or playing music can also be activities which are bought and sold, and this fact is an important part of the background to our everyday lives. The spheres of mutuality, equity and sharing are also those where activities are 'done for their own sake' rather than 'done for the sake of something else', in Aristotle's formulation (ibid.:

9). The relationship between the 'community' and 'market' is described by Gudeman as a dialectic or a constant tension, by which he means many things. They are opposed to each other, each cannot exist without the other, they overlap and borrow from each other (ibid.: 14, 95). The relationship evolves over time, and its evolution is one of the themes of the book. Gudeman says the relationship is unstable, but he clearly believes that something important has been lost in the last 30 years, during which the market has increasingly 'cascaded' and 'colonised' the realms of mutuality, permeating its spaces and squeezing out its values (ibid.: 148–65 et passim).

It is not always easy to disentangle what, for Gudeman, shapes the history of this relationship. In much of the discussion, the tension between these two realms plays out as though it was the fluctuating outcome of two different dispositions which all human beings possess. It is only in the most recent period that the realm of calculative reason is portrayed as dominant in the relationship, and more reference is made to agency, including those of international governance, the IMF, the World Bank and the WTO. This realm, which is variously described as that of competitive trade, impersonal trade, the market, becomes dominant with the emergence of a third tier: market finance, or the trade in money (ibid.: 14, 150; see also Gudeman 2010). The emphasis throughout is on exchange relations rather than production relations: capitalism as such is only fleetingly mentioned in the book, and not at all in the index.

Clearly there is an overlap between Gudeman's approach and the one we have developed, both in some of the ways the two realms are characterised, and in the insistence that they cannot be understood in isolation. However, there are two significant differences which explain why we think open and closed economies is a more appropriate framework for thinking about the kind of processes we have been describing.

In the social worlds of our case studies, the relationship between open and closed economies is not balanced, or one that can be described in terms of reciprocal influences. It is a relationship structured by dominance, and has been so for a long time, since before the triumph of the financial sphere. The appeal to the values of mutuality – for example, in supermarket sales techniques – does not make this a balanced relationship, anymore than the possible existence of mutuality in boardrooms and trading floors. Dispositions do not explain the relationship, or its evolution, though we would agree that the long-term development of specific economic institutions creates certain kinds of disposition and practice, to use Bourdieu's (1977) terminology. The open economy we have been describing includes trade and competition, and it also includes 'dispositions' towards calculative reason and self-interest, but they are not all of it. There is also the extraordinary power of a handful of corporations in the food chain, undermining the livelihoods of small farmers and shaping food choices. What farmers usually encounter is not competition but global corporations which have a monopoly position or behave as cartels.

These are the processes with which we started this book, while the ethnographic chapters documented resistance to this dominance, a resistance not to all uses of money or to all markets (though we have noted how these terms can become a shorthand for exploitation), but to the economic forces which extract value from their work and creativity. The attempt to increase autonomy through 'closure' is built around the values of mutuality, but it is also a political response to the power of corporations. That reaction focuses on the livelihood strategies of farmers and a reconfigured relationship with 'consumers', an alliance which is both crucial to increasing autonomy and difficult to stabilise. There are other political responses to this domination, which focus on labour relations in mainstream food chains or attacking the power of corporations. The two political strategies can coexist, and even the contributors to this book disagree about which might be most important in the long run.

A second advantage of the interpretation we have adopted is that open and closed, like autonomy, are relative terms. This sets up the relationship between the two spheres in a rather different way. When farmers and consumers talk about the kinds of social relations they want to establish, or the kinds of value they try to realise, we sense that they express this in terms of movement, of ambitions partially realised, so that both parts of the equation are co-present, though in different measures. One example is the attempt to develop a local food system, an ambition which only exists because in most places the open system has destroyed it. Each strategy we have documented pushes back against some part of the open system. Farmers reduce industrial inputs to increase self-sufficiency, new markets are created through direct sales. A further step is creating more permanent collective links between producers and consumers, attempting to recast that relationship. Each move is a position between open and closed, and that informs the understanding of what is going on. Then sometimes the fresh local foodstuff becomes a speciality, with organic or place-of-origin certification, and flies out of the closed economy.

Money, Values and Everyday Life

There are some objects, activities, even aspects of the world, which we think of as governed by precepts of sharing which are at odds with commercial logic – that is, with extracting profit from possession and exchange. Every time that line is crossed – as when selling the family silver – it triggers a conversation about morality. Recently in Britain we have seen examples of that moral conversation erupting dramatically at the time of reforms to the public sector. What is the appropriate role of profit-making corporations in the National Health Service? Is the primary purpose of higher education its contribution to the British economy or the creation of knowledge which should be free to all? In this last section we want to comment more generally on what happens when we move across the line between economic realms which are shaped by different principles and values. It will start with an example from one of our fieldwork sites, and involves the complex interaction between economic activity and aesthetic values. Beautiful things are famously both 'priceless' and can be worth a fortune. In this case, the beauty is attributed to a landscape, one which has generated a great deal of money.

The Tuscan farmers described in Chapter 4 live in the Val d'Orcia, an area which experienced sharp economic decline when the share-cropping system collapsed 50 years ago. The aristocrats who owned most of the estates sold up or went bankrupt, few new economic activities emerged and the overall population halved. It became a marginal or relictual space. Now it is 'an artistic, natural and cultural park', and a UNESCO World Heritage Site. All this happened at the moment when an unbridgeable gap opened up between the present and the past, that is with the way of life which generated the buildings and landscapes which need to be preserved. Heritage parks, like heritage tomatoes, are postmodern. The avenues of cypress trees are well tended, but if the wrong kind of houses are built, letters from intellectuals appear in the Rome newspapers. Planning restrictions have created protests and disquiet amongst local builders and farmers. As one of many who finds this landscape inspiring, Jeff Pratt does not want to side automatically with those who have been labelled philistines. However, something rather odd has happened. There was a long campaign to establish the unique historic and aesthetic value of this area, and then to prevent it being 'spoiled' by the wrong kind of economic activity. However, this campaign has generated a flood of new commercial opportunities: towns like San Quirico and Pienza, whose historic centres are not surprisingly intact, are now one long parade of boutique shops selling 'local' specialities: wines, oil, cheese, recipe books, paintings and photographs in every kind of packaging.

The example of a heritage site echoes the arguments we have been making about the certification of foodstuffs. Values which originate, or are claimed to originate, outside the market economy can be turned into money and appropriated by agents in the open economy. David Harvey offers one approach to this combination of aesthetic values and commercial opportunities. He is interested in the way particular income streams (monopoly rents) can be generated out of a 'tradable resource, commodity or location', which in some crucial respects is 'unique and non-replicable' (Harvey 2001: 395). There is a fine line to tread between emphasising the uniqueness of the commodity and its availability to the consumer, and the line is established through a prolonged series of cultural claims. If we look in detail at how a natural park gets established, or how a foodstuff obtains place-of-origin recognition, we find local authorities, entrepreneurs and a range of experts arguing over boundaries, traditions, qualities and by now familiar claims to authenticity. The same is famously true of art historians and critics, whose expertise can have such a dramatic effect on the price of a painting: Harvey's lecture was delivered at London's Tate Modern gallery.

We have encountered this process many times in the book, not least when discussing the importance of labels and certification. Like Harvey, we have looked at the ever-present possibility of commercial appropriation, and the response to that possibility. However, there is another more experiential aspect to all this. What happens, for example, when we look at a Picasso, having been instructed both that it is a great painting and that it is worth £50 million? Are those entirely independent facts which cannot be brought into the same frame? Or does the market value act as the signifier that it is a great painting, in which case why does it also signify that this is an investment which could be exchanged for 7000 tons of copper? For some people the equation is unproblematic. The rich frequently invest in expensive works of art, turning their piles of cash into something unique and non-replicable; it is a kind of 'trading-up', and they can even gain moral kudos in the process. A similar point is also made by Graeber (2012: 92–98).

Generally, movement across the line between the domains which we have called closed and open creates tensions. In the first domain there are social values in the sense used by Graeber (2001), and which we have explored throughout this book: the creative energy people invest in activities which are an end in themselves and which are embedded in social relations. In the second, value is measured by money and established through markets. Even if we do not buy works of art, there are many circumstances in which we find both these contrasting ways of understanding value present, and they create a moral tension which has to be negotiated.

We have seen that those buying food in farmers' markets often have to pay a price premium, especially in England. In other words, they have to pay the 'normal' price, which for most people is established by the supermarkets, plus an extra which represents an ethical commitment, or what is necessary to obtain quality. The extra covers those values of the food which originate outside commercial circuits, yet they are translated into a price and obtained through money. This means that the market for alternative foods is rather a hybrid place, where different principles meet and have somehow to be negotiated. The same is true for fair trade products and most forms of ethical consumption, and while we know that the appeal to ethics is increasing as a corporate marketing strategy, this does not explain everything that is going on. These two frames for recognising value are present in many everyday contexts, and in saying they are negotiated we are not implying that they are necessarily reconciled. Nor are we suggesting that we can generalise or easily predict what people do. Social positioning, cultural contexts, predispositions and other factors will influence actions; all we want to do is illustrate the range of responses with a few examples.

Growing your own fruit and vegetables on an allotment can save a significant sum of money. Like house repairs or making clothes, it was one of a number of activities undertaken by men and women 'after work' as an integral part of the household economy. The social contours of those who continue these activities, out of necessity or tradition, are less clear-cut than a generation ago, and there are certainly significant differences between the English and the Mediterranean case studies. In terms of growing food, sociological research has shown that there has always been more to allotment life than financial calculation (Crouch & Ward 1988). This has become more marked with changes in the social composition of those who 'grow their own'. In parts of Sussex there is a striking shift towards the professional middle classes and towards a more 'recreational' use of the plots. It is often the physicality of the work which people find rewarding. Unlike a day in front of the computer or a classroom of children, the work is objectified in a physical transformation: a bag of food and the sight of a freshly dug plot as you close the gate behind you. As with decorating a room or making clothes, there are practical skills built up over the long term, which people value and leave a physical legacy. Some people calculate the monetary value of the food they have grown, or the money saved by not employing a professional decorator. It may be an important reason for getting involved in these activities, or over time the skills acquired (from carpentry to cake-making) may themselves become a source of income. Or it may become a joke about how much a home-grown lettuce has cost, and jokes often reveal an underlying tension.

Other people, or the same people in other circumstances, try to keep the two realms completely separate. They do not want to think about this kind of activity in terms of money earned or saved because it 'devalues' the whole point of what they are doing. They will try to find ways to minimise cash outlays when growing food or doing up a house, and if they do have to spend money this is mentally ring-fenced and kept separate from the outcome. They are keen to keep a line between these activities, and feel some moral principle has been blurred if it is crossed. In effect, they are creating what have been called 'non-commoditised spaces' in their everyday lives, small closures in an open system, where different kinds of values and social relations can flourish. For some, these spaces are more deliberately constructed as a way of building critical awareness, which can become what has been called a 'hybrid strategy' (Leahy 2011) of resistance to dominant economic forces. At the most politicised and 'purist' end of this spectrum, at least in terms of our examples, are the 'de-growth' activists of the Tarn.

But why do some people think a moral principle has been blurred if money enters into the equation? After all, Miller has argued that creativity often comes from linking these two worlds, the alienable which has a price and the inalienable which is priceless: 'value is most effectively created by its own use as a bridge between what otherwise would be regarded as distinct regimes of value' (Miller 2008: 1130). Much discussion of this issue in this book and elsewhere focuses on the way money makes two objects commensurable: the home-grown lettuce is just like the one bought in a shop. But if we go back to the 'moral conversations' triggered when a line is crossed, we see that it is often not about things but about relationships. This clarifies why moral arguments emerge so strongly in these conversations. When farmers talk about the dilemmas of setting a price for the goods they sell, they are wrestling with the interaction of two economic circuits, each governed by very different principles. Some try to keep them separate because they prioritise, or would like to prioritise, various forms of mutuality. Some can see financial advantage (in the form of price premiums or customer loyalty) of linking these realms of value. And some make their decisions for an evolving mixture of reasons, because life is messy and in any case they are in a subordinate position in the economic scheme of things.

When people obtain their food from different sources, some from the mainstream commercial circuits of the supermarket, some from outside it, their choices may not be very consistent, but nor are they entirely random. What we called a dual food economy is another way in which people can separate out different economic circuits: some food is more 'short-term' and utilitarian, some more associated with long-term patterns of mutuality, for example creating meals which mark events and create social boundaries through commensality. These values can coexist with less egalitarian concerns, such as demonstrations of cultural capital and the status associated with hospitality, something which emerges at various points in the ethnography. The intention here is not to analyse any of these issues in any more detail, but to suggest that the experience and the voices of those we interviewed for this book can illuminate the complex negotiations we all make in our everyday lives.

10

Afterword

In this book we have explored the everyday reality of those who are creating more closed economies in order to achieve greater autonomy and realise the values associated with mutuality. Those who create these movements reject the dominant values of unfettered market relations, and in many cases have themselves been rejected or ejected from the mainstream economy. Here there is an overlap with a much wider phenomenon, one that is urban more than rural, and has accelerated dramatically since the onset of the financial crisis that commenced in 2008. We refer to the millions of people who have become unemployed or underemployed at a time when austerity measures have substantially reduced state provision for the poorest. This extended crisis is transforming economic life dramatically, not least in southern European, and affecting young adults in particular. The farmers in our book have always lived with fluctuating incomes and a great deal of economic insecurity, while some have made a deliberate choice to reject consumerism and monetary values. The on-going crisis has increasingly hit families whose economic life has been based on a regular wage and the dull economic compulsion of bills, housing costs and other unavoidable expenditure. Does our account of values, and strategies of closure, throw any light on this wider situation?

We know the raw statistics of this situation, and some of the survival strategies of those hardest hit. People sometimes find work in the 'black' economy, usually on a precarious day-to day basis and on very low rates of pay. Wider kin groups pool their income and resources, whether food or a place to sleep. Adult children move back home, sometimes whole families return to rural districts where grandparents can offer access to land and accommodation which is free of rent or mortgage payments. Goods are exchanged and bartered, scavenged and recycled. Sometimes property rights are challenged more directly, buildings and farmland appropriated, supermarkets raided and their goods redistributed to food kitchens. In our own research in Andalusia and Tuscany we witnessed many of these things happening, as small towns absorbed those thrown out of the construction and tourist industries, or family businesses were bankrupted in the recession.

There are a number of books which analyse the work of mobile, highly networked activists who are using their skills to stimulate forms of direct democracy (e.g. Castells 2012; Graeber 2013; Maeckelbergh 2009). If they are the spark trying to ignite new forms of social action, we know less about the highly varied landscape in which they operate. Variations derive from the strength and geographical concentration of kinship networks, and the kinds of solidarity which have long roots in neighbourhoods and workplaces. The social landscape also includes the actions and priorities of local councils, churches, social centres, cooperatives and credit unions. In turn these reflect the strength of long-standing political cultures. Not all of them are part of the traditional Left. Opposition to migrants surfaces frequently in everyday conversations, and is broadcast by xenophobic populist parties which in some regions have developed their own rooted social movements.

One recent study starts to fill the gap in our knowledge by giving us some details about the rapid growth of alternative economic practices in Barcelona (Conill et al. 2012). The researchers used a survey to reveal the range of formally constituted networks and organisations, as well as those practices of sharing and non-monetised exchanges which are embedded in everyday life. They used focus groups and questionnaires to examine the range of people involved in them, and generated a number of categories to describe levels of participation. The key distinction they make is between those who are 'rooted in the quest for the use value of life' (ibid.: 211) as opposed to those who are still to some extent locked into expectations based on increasing consumption. This picks up a theme in Chapter 1, that we are living through 'a cultural crisis, of non-sustainability of certain values as the guiding principle of human behaviour' (Castells et al. 2012: 13).

A number of points stand out. Barcelona has a well-developed sector of ethical banking and financial cooperatives; after that, by far the most important organisations, accounting for three quarters of all activity, involve food. These are either agro-ecological producers or consumer groups. A few are producer–consumer cooperatives, and are linked to Andalusian associations such as La Ortiga in Seville, which we mentioned in Chapter 6. This suggests that the kind of food movements examined in our book are a crucial component of an emerging alternative economy. A second point is that the Barcelona networks and organisations were founded before the onset of the crisis, by people who the researchers designate as 'culturally transformative' (Conill et al. 2012: 221); that is, they are consciously and actively trying to create an economy based on alternative values. Alongside the active transformers, the study uses two other categories to explore this evolving economy: those who have begun to participate in these activities since the crisis began ('practitioners'), and those who are still rooted in the cultural assumptions of the pre-crisis economy. The social profile of each of these three categories is complex and in many ways incomplete, but the rough pattern which emerges from the data suggests a final issue.

The more active innovators tend to be younger, better educated, and often have skilled jobs or flexible working hours. They created networks and organisations before the recession out of long-standing explicit opposition to the values of capitalism, and many are now part of the *indignados* occupying city squares. At the other extreme, those who are retired (a quarter of the city's population) are the least involved. This is unlikely to be because they are hooked into particularly affluent patterns of consumption, or because they opt out of the everyday practices of sharing, exchange and mutuality. Instead, as the authors suggest, it is because they (and many other parts of the city's population) are embedded in a culture of work which has given meaning to their lives (ibid.: 225). They have, or had, a regular wage and a set of largely predictable monetary outgoings; over the long term, they experienced a rising material standard of living which provides much of the narrative shape of their lives. The crisis has smashed these expectations, not just for the older generations but for the majority of people in the city. The response to this crisis from the mainstream Left is to restore these aspirations through getting the open economy to function in a more regulated and egalitarian manner. Conversely, the radical transformers are seeking to expand the values and practices of what we have termed a closed economy until they replace the existing system entirely. That, at least, is one way of framing the political tensions of the moment.

References

Agence BIO (2013) Website at www.agencebio.org, accessed 15 April 2013.

Andrews, G. (2008) *The Slow Food Story* (London: Verso).

ARS (n.d.) 'Food and Local Products', Action in Rural Sussex. Available at: www.ruralsussex.org.uk/service/food-local-products/, accessed 10 March 2013.

Autoría Colectiva (2006) *Las Pies en la Tierra: reflexiones y experiencias hacia un movimiento agroecológico* (Barcelona: Virus Editorial).

Baqué, P. (2012) *La Bio entre business et projet de société* (Marseilles: Edition Agone).

Barham, E. (2003) 'Translating *Terroir* : The Global Challenge of French AOC Labelling', *Journal of Rural Studies* 19: 127–138.

Baykan, B.G. (2007) 'From Limits to Growth to Degrowth within French Green Politics', *Environmental Politics* 16 (3): 513–517.

Beckert, J. (2009) 'The Great Transformation of Embeddedness: Karl Polanyi and the New Economic Sociology', in C. Hann and K. Hart (eds), *Market and Society: The Great Transformation Today*, pp. 38–55 (Cambridge: Cambridge University Press).

BHFP (2012) 'The View from a Mountain: An Organic Food Grower's Story', Brighton and Hove Food Partnership. Available at: www.bhfood.org. uk/blogs/entry/the-view-from-a-mountain-an-organic-food-grower-s-perpective, accessed 10 April 2013.

BHO (2012) 'The Borough of Lewes: Parliamentary, Economic and Religious History', British History Online. Available at: www.british-history.ac.uk/ report.aspx?compid=56910, accessed 15 March 2013.

Bourdieu, P. (1977) *Outline of a Theory of Practice* (Cambridge: Cambridge University Press).

—— (1984) *Distinction: A Social Critique of the Judgement of Taste* (London: Routledge).

Bové, J., and F. Dufour (2001) *The World Is Not for Sale* (London: Verso).

Brenan, G. (1990) *The Spanish Labyrinth: An Account of the Social and Political Background of the Spanish Civil War* (Cambridge: Cambridge University Press).

Briault, E. (1942) 'Sussex (East and West), Parts 83–84', in L.D. Stamp (ed.), *The Land of Britain: The Report of the Land Utilisation Survey of Britain*, Vol. 8: *South-eastern England*, pp. 471–555 (London: Geographical Publications).

Buck, D., C. Getz and J. Guthman (1997) 'From Farm to Table: The Organic Vegetable Commodity Chain of Northern California', *Sociologia Ruralis* 37 (1): 3–20.

Burchardt, J. (2002) *Paradise Lost: Rural Idyll and Social Change since 1800* (London: I.B. Taurus).

Burr, A. (1999) 'Can Farmers' Markets Improve Access to Fresh Local Produce for Families on Low Incomes? A Report of a Participatory Investigation Held in Conjunction with Pilot Farmers' Markets in Lewes, East Sussex'. Available at: www.commoncause.org.uk/pdfs/CanFarmersMarketsImproveAccess. pdf, accessed 7 March 2013.

Burr, A.M., T. Jewell and K. Rayner (1999) 'Sussex Farmers' Market: An Evaluation of Three Pilot Markets in Lewes'. Available at: www.common cause.org.uk/pdfs/FMEvaluationReport1298.pdf, accessed 7 March 2013.

Camporesi, P. (1993) *The Magic Harvest* (Cambridge: Polity Press).

Caplan, P. (2000) 'Eating British Beef with Confidence: Perceptions of the Risk of BSE in London and West Wales', in P. Caplan (ed.), *Risk Revisited*, pp. 184–203 (London: Pluto Press).

Carrier, J. (1995) *Gifts and Commodites: Exchange and Western Capitalism since 1700* (London: Routledge).

—— (2012) 'Introduction', in J. Carrier and P. Luetchford (eds), *Ethical Consumption: Social Value and Economic Practice*, pp. 1–36 (Oxford: Berghahn).

Carrier, J., and P. Luetchford (eds) (2012) *Ethical Consumption: Social Value and Economic Practice* (Oxford: Berghahn).

Carrier, J., and R. Wilk (2012) 'Conclusion', in J. Carrier and P. Luetchford (eds), *Ethical Consumption: Social Value and Economic Practice*, pp. 217–228 (Oxford: Berghahn).

Carsten, J. (1997) *The Heat of the Hearth: The Process of Kinship in a Malay Fishing Community* (Oxford: Clarendon Press).

Castells, M. (2012) *Networks of Outrage and Hope* (Cambridge: Polity Press).

Castells, M., J. Caraça and G. Cardoso (eds) (2012) *Aftermath: The Cultures of the Economic Crisis* (Oxford: Oxford University Press).

Clemente, P. (1987) *Il Mondo a Meta: sondaggi antropologici sulla Mezzadria classica* (Rome: Istituto Alcide Cervi, Editrice Mulino).

Conford, P. (2001) *The Origins of the Organic Movement* (Edinburgh: Floris).

Conill, J., M. Castells, A. Cardenas and L. Servon (2012) 'Beyond the Crisis: The Emergence of Alternative Economic Practices', in M. Castells, J. Caraça, and G. Cardoso (eds) *Aftermath: The Cultures of the Economic Crisis*, pp. 210–248 (Oxford: Oxford University Press).

Corbin, J. (1993) *The Anarchist Passion: Class Conflict in Spain 1810–1965* (Aldershot: Avebury).

Cottingham, M., and E. Winkler (2007) 'The Organic Consumer', in S. Wright and D. McCrea (eds), *The Handbook of Organic and Fair Trade Marketing*, pp. 29–52 (Oxford: Blackwell).

Counihan, C. (1984) 'Bread as World: Food Habits and Social Relations in Modernizing Sardinia', *Anthropological Quarterly* 57 (2): 47–59.

CPRE (2012) From Field to Fork: The Value of England's Local Food Webs', Campaign to Protect Rural England. Available at: http://www.cpre.org. uk/resources/farming-and-food/local-foods/item/2897-from-field-to-fork, accessed 8 March 2013.

Crouch, D., and C. Ward (1988) *The Allotment: Its Landscape and Culture* (London: Faber and Faber).

Cuéllar Padilla, M. (2010) 'La Certificación Ecológica como Instrumento de Revalorización de lo Local: Los Sistemas Participativos de Garantía en Andalucía', in M. Soler Montiel and C. Guerrero Quintero (eds), *Patrimonio*

Cultural en la Nueva Ruralidad Andaluza, pp. 285–294 (Seville: Junta de Andalucia).

Dale, G. (2010) *Karl Polanyi: The Limits of the Market* (Cambridge: Polity Press).

DEFRA (2011) 'Food Transport Indicators to 2009/10', Department for Environment, Food and Rural Affairs, UK. Available at: www.defra.gov.uk/statistics/files/defra-stats-foodfarm-food-transport-statsnotice-110331.pdf, accessed 20 March 2013.

Del Campo Trejedor, A. (2000) *Agricultores y Ganaderos Ecológicos en Andalucia* (Seville: Junta de Andalucia).

Deléage, E. (2004) *Paysans de la Parcelle à la Planète: socio-anthropologie du réseau agriculture durable* (Paris: Editions Syllepse).

De Neve, G., P. Luetchford and J. Pratt (2008) 'Introduction: Revealing the Hidden Hands of Global Market Exchange', in G. de Neve, P. Luetchford, J. Pratt and D. Wood (eds), *Hidden Hands in the Market: Ethnographies of Fair Trade, Ethical Consumption and Corporate Social Responsibility*, pp. 1–30 (Bingley: Emerald).

De Neve, G., P. Luetchford, J. Pratt and D. Wood (eds) (2008) *Hidden Hands in the Market: Ethnographies of Fair Trade, Ethical Consumption and Corporate Social Responsibility* (Bingley: Emerald).

Dickie, J. (2007) *Delizia* (London: Sceptre).

Donaire, G. (2011) 'Fin a la Discriminación en el Campo', *El Pais*, 7 December. Available at: www.elpais.com/articulo/andalucia/Fin/discriminacion/campo, accessed 29 November 2012.

Douglas, M. (1975) 'Deciphering a Meal', in *Implicit Meanings: Selected Essays in Anthropology*, pp. 249–275 (London: Routledge).

Du Puis, M. (2000) 'Not in My Body: rBGH and the Rise of Organic Milk', *Agriculture and Human Values* 17: 285–295.

Du Puis, M., and D. Goodman (2005) 'Should We Go "Home" to Eat? Towards a Reflexive Politics of Localism', *Journal of Rural Studies* 21: 359–371.

Edelman, M. (2005) 'When Networks Don't Work: The Rise and Fall of Civil Society Initiatives in Latin America', in J. Nash (ed.), *Social Movements: An Anthropological Reader*, pp. 29–45 (London: Blackwell).

Eurostat (2010) 'Agricultural Labour Input'. Available at: http://epp.eurostat.ec.europa.eu/statistics_explained/index.php/Agricultural_labour_input, accessed 12 March 2013.

Fine, B. (2002) *The World of Consumption: The Material and Cultural Revisited* (London: Routledge).

Firth, R. (1975) 'The Sceptical Anthropologist? Social Anthropology and Marxist Views on Society', in M. Bloch (ed.), *Marxist Analyses and Social Anthropology*, pp. 29–60 (London: Malaby Press).

Fischler, C. (1980) 'Food Habits, Social Change and the Nature/Culture Dilemma', *Social Science Information* 19: 937–953.

Foucault, M. (1991) 'Governmentality', in G. Burchell, C. Gordon and P. Miller (eds), *The Foucault Effect: Studies in Governmentality*, pp. 87–104 (Chicago: University of Chicago Press).

Fournier, V. (2008) 'Escaping from the Economy: The Politics of Degrowth', *International Journal of Sociology and Social Policy* 28: 528–545.

Foweraker, J. (1989) *Making Democracy in Spain: Grass-roots Struggle in the South 1955–1975* (Cambridge: Cambridge University Press).

FPN (2012) 'Face the Difference: The Impact of Low Pay in the National Supermarket Chains', Fair Pay Network. Available at: www.fairpaynetwork. org, accessed April 16 2013.

Frow, J. (1997) *Time and Commodity Culture: Essays in Cultural Theory and Postmodernity* (Oxford: Clarendon Press).

Gibbon, P. (2008) 'An Analysis of Standards-based Regulation in the EU Organic Sector, 1991–2007', *Journal of Agrarian Change* 8 (4): 553–582.

Giorgetti, G. (1982) *Le Crete Senesi nell'Eta Moderna* (Florence: Olshki Editore).

Godelier, M. (1998) *The Enigma of the Gift* (Cambridge: Polity Press).

Godfrey, J. (2002), 'Land Ownership and Farming on the South Downs in West Sussex c. 1840–1940', *Sussex Archeological Collections* 140: 113–123.

Goodman, D., E.M. Du Puis and M. Goodman (2012) *Alternative Food Networks: Knowledge, Practice and Politics* (London: Routledge).

Goodman, D., and M. Redclift (1991) *Refashioning Nature: Food, Ecology and Culture* (London: Routledge).

Goodman, D., B. Sorj and J. Wilkinson (1987) *From Farming to Biotechnology* (Oxford: Blackwell).

Goody, J. (1982) 'Industrial Food: Towards the Development of a World Cuisine', in *Cooking, Cuisine and Class: A Study in Comparative Sociology*, pp. 154–174 (Cambridge: Cambridge University Press).

Graeber, D. (2001) *Towards an Anthropological Theory of Value: The False Coin of Our Own Dreams* (New York: Palgrave).

—— (2011) *Debt: The First 5000 Years* (New York: Melville House Publishing).

—— (2012) *Revolutions in Reverse* (London: Minor Compositions).

—— (2013) *The Democracy Project* (London: Allen Lane).

Grasseni, C. (2003) 'Packaging Skills: Calibrating Cheese to the Global Market', in S. Strasser (ed.), *Commodifying Everything*, pp. 259–288 (New York: Routledge).

Gregory, C. (1982) *Gifts and Commodities* (New York: Academic Press).

Gudeman, S. (1986) *Economics as Culture: Models and Metaphors of Livelihood* (London: Routledge).

—— (2001) *The Anthropology of Economy: Community, Market and Culture* (Oxford: Blackwell).

—— (2008) *Economy's Tension: The Dialectics of Community and Market* (Oxford: Berghahn).

—— (2010) 'Creative Destruction', *Anthropology Today* 26 (1): 3–7.

Gudeman, S., and A. Rivera (1990) *Conversations in Colombia: The Domestic Economy in Life and Text* (Cambridge: Cambridge University Press).

Guerrero Quintero, C., and M. Soler Montiel (2010) *Patrimonio Cultural en la Nueva Ruralidad Andaluza* (Seville: Junta de Andalucia).

Guthman, J. (2004a) *Agrarian Dreams: The Paradox of Organic Farming in California* (Berkeley: University of California Press).

—— (2004b) 'The Trouble with "Organic-lite" in California: A Rejoinder to the Conventionalization Debate', *Sociologia Ruralis* 44 (3): 301–320.

Hann, C., and K. Hart (eds) (2009) *Market and Society: The Great Transformation Today* (Cambridge: Cambridge University Press).

Hart, K. (1986) '"Heads or Tails?" Two Sides of the Coin', *Man* 21 (3): 637–656.

Harvey, D. (2001) 'The Art of Rent: Globalisation and the Commodification of Culture' in *Spaces of Capital*, pp. 394–411 (Edinburgh: Edinburgh University Press).

—— (2012) *Rebel Cities* (London: Verso).

Hinrichs, C. (2000) 'Embeddedness and Local Food Systems: Notes on Two Types of Direct Agricultural Market', *Journal of Rural Studies* 16 (3): 295–303.

Howkins, A. (2003) *The Death of Rural England: A Social History of the Countryside Since 1900* (London: Routledge).

Hughes, D. (2011) 'Rural Labour: Photographic Representation of the English Countryside in the 1930s Socialist Press', *History of Photography* 35 (1): 59–75.

IME (2012) 'Global Food Report', Institution of Mechanical Engineers. Available at: www.imeche.org/knowledge/themes/environment/global-food, accessed 12 February 2012.

Junta de Andalucia (n.d.) *II Plan Andaluz de Agricultura Ecológica (2007–2013)* (Seville: Consejería de Agricultura y Pesca, Junta de Andalucia).

Kahn, J. (1995) *Culture, Multiculture, Postculture* (London: Sage).

—— (1997) 'Demons, Commodities and the History of Anthropology', in J. Carrier (ed.), *Meanings of the Market: The Free Market in Western Culture*, pp. 69–98 (Oxford: Berg).

Kaplan, T. (1977) *Anarchists of Andalusia 1863–1903* (Princeton: Princeton University Press).

Keenan, J. (2012), 'Sussex Food and Drink Suppliers Call For Public Sector Support', *The Argus*, 12 June. Available at www.theargus.co.uk/news/business/9755469.Sussex_food_and_drink_suppliers_call_for_public_sector_support/, last accessed 7 March 2013.

Kneafsey, M., R. Cox, L. Holloway, E. Dowler, L. Venn and N. Toumaine (2008) *Reconnecting Consumers, Producers and Food: Exploring Alternatives* (Oxford: Berg).

Kopytoff, I. (1986) 'The Cultural Biography of Things: Commoditization as Process', in A. Appadurai (ed.), *The Social Life of Things: Commodities in Cultural Perspective*, pp. 64–94 (Cambridge: Cambridge University Press).

Lappé, F. (2013) 'Beyond the Scarcity Scare: Reframing the Discourse of Hunger with an Eco-mind', *Journal of Peasant Studies* 40 (1): 219–238.

Lawrence, F. (2004) *Not on the Label* (London: Penguin).

—— (2008) *Eat Your Heart Out* (London: Penguin).

Leahy, T. (2011) 'The Gift Economy', in A. Nelson and F. Timmerman (eds), *Life without Money*, pp. 111–135 (London: Pluto Press).

Leger, D., and B. Hervieu (1979) *Le Retour à la Nature: Au Fond de la Forêt … l'Etat* (Paris: Seuil).

Leger-Hervieu, D., and B. Hervieu (1983) *Des Communautés pour les Temps Difficiles: Néo-ruraux ou Nouveaux Moines* (Paris: Le Centurion).

Leroux, B. (2011) 'Les Agriculteurs Biologiques et l'Alternative: Contribution à l'Anthropologie Politique d'un Monde Paysan en Devenir', unpublished PhD thesis (Paris: CSE, Ecole des Hautes Etudes en Sciences Sociales).

Lewis, T., and E. Potter (eds) (2011) *Ethical Consumption: A Critical Introduction* (London: Routledge).

LFM (2013a) 'Stallholder Guidelines', Lewes Food Market. Available at: www.lewesfoodmarket.co.uk/stallholder_guidelines, accessed 25 March 2013.

—— (2013b) 'Home Page', Lewes Food Market. Available at: www.lewesfoodmarket.co.uk, accessed 28 March 2013.

Lie, J. (1991) 'Embedding Polanyi's Market Society', *Sociological Perspectives* 34 (2): 219–235.

Littler, J. (2011) 'What's Wrong with Ethical Consumption?' in T. Lewis and E. Potter (eds), *Ethical Consumption: A Critical Introduction*, pp. 27–39 (London: Routledge).

Lockie, S., and D. Halpin (2005) 'The Conventionalisation Thesis Reconsidered: Structural and Ideological Transformation of Australian Organic Agriculture', *Sociologia Ruralis* 45 (4): 284–307.

López Garcia, D., and A. López López (2003) *Con la Comida no se Juega: alternativas autogestionarias a la globalización capitalista desde la agroecología y el consumo* (Madrid: Traficante de Sueños).

Lotti, A. (2010) 'The Commoditization of Products and Taste: Slow Food and the Conservation of Agrobiodiversity', *Agriculture and Human Values* 27: 71–83.

Luetchford, P. (2008a) *Fair Trade and a Global Commodity: Coffee in Costa Rica* (London: Pluto).

—— (2008b) 'Hidden Hands in Fair Trade: Beyond the Charms of the Family Farm', in G. de Neve, P. Luetchford, J. Pratt, and D. Wood (eds), *Hidden Hands in the Market: Ethnographies of Fair Trade, Ethical Consumption and Corporate Social Responsibility*, pp. 143–169 (Bingley: Emerald).

—— (2012a) 'Consuming Producers: Fair Trade and Small Farmers', in J. Carrier and P. Luetchford (eds), *Ethical Consumption: Social Value and Economic Practice*, pp. 60–80 (Oxford: Berghahn).

—— (2012b) 'Economic Anthropology and Ethics', in J. Carrier (ed.), *A Handbook of Economic Anthropology*, 2nd edn., pp. 397–411 (Cheltenham: Edward Elgar).

Luetchford P., and J. Pratt (2011) 'Values and Markets: An Analysis of Organic Farming Initiatives in Andalusia', *Journal of Agrarian Change* 11 (1): 87–103.

Luetchford, P., J. Pratt and M. Soler Montiel (2010) 'Struggling for Autonomy: From Estate Labourers to Organic Farmers in Andalusia', *Critique of Anthropology* 30 (3): 313–321.

Maeckelbergh, M. (2009) *The Will of the Many: How the Alterglobalization Movement Is Changing the Face of Democracy* (London: Pluto).

Malefakis, E. (1970) *Agrarian Reform and Peasant Revolution in Spain* (New Haven: Yale University Press).

Malinowski, B. (2007[1922]) *Argonauts of the Western Pacific* (London: Routledge).

Martinez-Alier, J. (1971) *Labourers and Landowners in Southern Spain* (London: Allen and Unwin).

Mauss, M. (2002[1925]) *The Gift: The Form and Reason for Exchange in Archaic Societies*, trans. W.D. Halls (London: Routledge).

May, E. (1997) '"Primitive Rebels" in Spain: Historians and the Anarchist Phenomenon', in R. Stradling, S. Newton and D. Bates (eds), *Conflict and Coexistence: Nationalism and Democracy in Modern Europe*, pp. 196–218 (Cardiff: University of Wales Press).

Mazower, M. (1998) *Dark Continent* (London: Penguin).

Michelson, J. (2001) 'Recent Development and Political Acceptance of Organic Farming in Europe', *Sociologia Ruralis* 41 (1): 3–20.

Miller, D. (1995) 'Introduction: Anthropology, Modernity and Consumption', in D. Miller (ed.), *Worlds Apart: Modernity through the Prism of the Local*, pp. 1–22 (London: Routledge).

—— (1998) *A Theory of Shopping* (Cambridge: Polity Press).

—— (2008) 'The Uses of Value', *Geoforum* 39: 1122–1132.

Mintz, J. (1982) *The Anarchists of Casas Viejas* (Chicago: University of Chicago Press).

MLFW (2008) 'CPRE: Food Webs and Mapping', Making Local Food Work. Available at: www.makinglocalfoodwork.co.uk/who/CPREfoodwebsand mapping.cfm, accessed 15th March 2013.

Morgan, K., T. Marsden and J. Murdoch (2006) *Worlds of Food: Place, Power and Provenance in the Food Chain* (Oxford: Oxford University Press).

Nelson, A., and F. Timmerman (2011) *Life without Money: Building Fair and Sustainable Economies* (London: Pluto).

NHS (2012) 'Obesity', National Health Service. Available at: www.nhs.uk/ conditions/Obesity/Pages/Introduction, accessed 25 March 2013.

Okura Gagné, N. (2011) 'Eating Local in a US City: Reconstructing "Community" – a Third Place – in a Global Neo-liberal Economy', *American Ethnologist* 38 (2): 281–293.

Ortner, S. (1984) 'Theory in Anthropology since the Sixties', *Comparative Studies in Society and History* 26 (1): 126–166.

Parry, J. (1986) 'The Gift, the Indian Gift and the "Indian Gift"', *Man* 21 (3): 453–473.

Parry, J., and M. Bloch (eds) (1989) *Money and the Morality of Exchange* (Cambridge: Cambridge University Press).

Patel, R. (2007) *Stuffed and Starved* (London: Portobello Books).

Pazzagli, C. (1979) *Per la Storia dell'Agricoltura Toscana nei Secoli XIX e XX* (Turin: Einaudi)

Petrini, C. (2001) *Slow Food: The Case for Taste* (New York: Columbia University Press).

Pinkerton, T., and R. Hopkins (2009) *Local Food: How to Make it Happen in Your Community* (Totnes: Transition Books).

Ploeg, J. van der (2009) *The New Peasantries* (London: Earthscan).

—— (2010) 'The Peasantries of the Twenty-first Century: The Commoditisation Debate Revisited', *Journal of Peasant Studies* 37 (1): 1–30.

Polanyi, K. (1957) 'The Economy as Instituted Process', in K. Polanyi, C. Arensberg and H. Pearson (eds), *Trade and Market in the Early Empires: Economies in History and Theory*, pp. 243–270 (New York: Free Press).

—— (2001[1944]) *The Great Transformation: The Political and Economic Origins of Our Time* (Boston: Beacon Press).

Pollan, M. (2006) *The Omnivore's Dilemma* (London: Penguin).

Poole, S. (2011) 'Let's End Our Obsession with Making Food Sexy', *Observer*, 11 December.

Pouzenc, M., and J. Pilleboue (2009) 'AMAP dans l'alimentation, une nouvelle forme de rapports consommateurs–producteurs?' Available at: www.cafe-geo.net/article.php3?id_article=1565, accessed 30 June 2012.

Pratt, J. (1987) 'Time, Work and Money in a Tuscan Co-operative', *Journal of Peasant Studies* 14 (4): 435–453.

—— (1994) *The Rationality of Rural Life* (Chur, Switzerland: Harwood Academic Press).

—— (2003) *Class, Nation and Identity* (London: Pluto)

—— (2007) 'Food Values: The Local and the Authentic', *Critique of Anthropology* 27: 285–300.

—— (2009) 'Incorporation and Resistance: Analytical Issues in the Conventionalization Debate and Alternative Food Chains', *Journal of Agrarian Change* 9 (2): 155–174.

Proust, M. (2006) *Remembrance of Things Past*, Vol. 1, trans. C.K. Scott Moncrieff (Ware: Wordsworth Editions).

Quellier, F. (2007) *La table des Français: Une histoire culturelle, XVe-début XIXe siècle* (Rennes: Presses Universitaires de Rennes).

Reed, M. (1984) 'The Peasantry of Nineteenth-century England: A Neglected Class?' *History Workshop Journal* 18: 53–76.

Robbins, J. (2009) 'Rethinking Gifts and Commodities: Reciprocity. Recognition and the Morality of Exchange', in K. Browne and L.B. Milgram (eds), *Economics and Morality: Anthropological Approaches*, pp. 43–58 (Lanham, MD: Altamira Press).

Roberts, P. (2008) *The End of Food* (London: Bloomsbury).

Romero, F. (2003) *Historia de Puerto Serrano: Puerto Serrano Contemporaneo* (Cadiz: Servicio de Publicaciones, Diputatión de Cádiz).

RSA (2010) 'RSA Animate – First as Tragedy, Then as Farce'. Available at: http://www.youtube.com/watch?v=hpAMbpQ8J7g, accessed 29 April 2013.

Sahlins, M. (1974) *Stone Age Economics* (London: Tavistock).

Salmona, M. (1994) *Souffrances et résistances des paysans français: violences des politiques publiques de modernisation économique et culturelle* (Paris: L'Harmattan).

Sevilla Guzmán, E., and A. Alonso Mielgo (2005) 'Entre la agroecología, como movimiento social, y la agricultura orgánica, como negocio: el caso de las asociaciones andaluzas de productores-consumidores', *Almirez* 13: 337–387.

Silverman, S. (1975) *The Three Bells of Civilisation* (New York: Columbia University Press).

Simonetti, L. (2010) *Mangi, chi può. Meglio, meno e piano: L'ideologia di Slow Food* (Firenze: Mauro Pagliai Editore).

Sobo, E. (1994) 'The Sweetness of Fat: Health, Procreation, and Sociability in Rural Jamaica', in N. Sault (ed.), *Many Mirrors: Body Image and Social Relations*, pp. 132–154 (New Brunswick, NJ: Rutgers University Press).

Steel, C. (2008) *Hungry City: How Food Shapes Our Lives* (London: Chatto and Windus).

Strathern, M. (1988) *The Gender of the Gift* (Berkeley: University of California Press).

Strong, R. (2003) *Feast: A History of Grand Eating* (London: Pimlico).

Stuart, T. (2009) *Waste: Uncovering the Global Food Scandal* (London: Penguin).

Sustainweb (2013) 'Sustain the Alliance for Better Food and Farming'. Available at: www.sustainweb.org, accessed 7 March 2013.

Sutton, D. (2001) *Remembrance of Repasts: An Anthropology of Food and Memory* (Oxford: Berg).

Taillefer, F. (1978) *Atlas et géographie du midi toulousain* (Paris: Flammarion).

Taylor, M., D. Carrington and F. Lawrence (2013) 'Horsemeat Scandal', *Guardian*, 16 February.

TLP (2013) 'What It Is', The Lewes Pound. Available at: www.thelewespound. org/what-it-is/, accessed 20 April 2013.

Touraine, A., and F. Dubet (1981) *Le Pays contre L'Etat* (Paris: Seuil).

Treagar, A. (2003) 'From Stilton to Vimto: Using Food History to Re-think Typical Products in Rural Development', *Sociologia Ruralis* 43 (2): 91–107.

Tudge, C. (2007) *Feeding People Is Easy* (Grosseto: Pari Publishing).

Van Dam, D., J. Nizet, M. Dejardin and M. Streith (2009) *Les agriculteurs biologiques: ruptures et innovation* (Dijon: Educagri).

Varul, M. (2008) 'Consuming the Campesino: Fair-trade Marketing Between Recognition and Romantic Commodification', *Cultural Studies* 22 (5): 654–679.

Veronelli, L., and P. Echaurren (2003) *Le Parole della Terra* (Viterbo: Nuovi Equilibri).

Warde, A. (1997) *Consumption, Food and Taste* (London: Sage).

Weiner, A. (1992) *Inalienable Possessions: The Paradox of Keeping while Giving* (Berkeley: University of California Press).

Wilk, R. (1995) 'Learning to Be Local in Belize: Global Systems of Common Difference', in D. Miller (ed.), *Worlds Apart: Modernity through the Prism of the Local*, pp. 110–133 (London: Routledge).

—— (2006) 'From Wild Weeds to Artisanal Cheese', in R. Wilk (ed.), *Fast Food/Slow Food: The Cultural Economy of the Global Food System*, pp. 13–27 (Lanham, MD: Altamira Press).

Williams, G. (2008a) 'Cultivating Autonomy: Power, Resistance and the French Alterglobalization Movement', *Critique of Anthropology* 28 (1): 63–86.

—— (2008b) *Struggles for an Alternative Globalization: An Ethnography of Counterpower in Southern France* (Aldershot: Ashgate).

Williams, R. (1973) *The Country and the City* (London: Chatto and Windus).

Wright, C. (2004) 'Consuming Lives, Consuming Landscapes: Interpreting Advertisements for Cafédirect Coffees', *Journal of International Development* 16 (5): 665–680.

Zizek, S. (2009) *First as Tragedy, Then as Farce* (London: Verso).

Index

Compiled by Sue Carlton